Multimodal Biometric Identification System

This book presents a novel method of multimodal biometric fusion using a random selection of biometrics, which covers a new method of feature extraction, a new framework of sensor-level and feature-level fusion. Most of the biometric systems presently use unimodal systems, which have several limitations. Multimodal systems can increase the matching accuracy of a recognition system. This monograph shows how the problems of unimodal systems can be dealt with efficiently, and focuses on multimodal biometric identification and sensor-level, feature-level fusion. It discusses fusion in biometric systems to improve performance.

- Presents a random selection of biometrics to ensure that the system is interacting with a live user.
- Offers a compilation of all techniques used for unimodal as well as multimodal biometric identification systems, elaborated with required justification and interpretation with case studies, suitable figures, tables, graphs, and so on.
- Shows that for feature-level fusion using contourlet transform features with LDA for dimension reduction attains more accuracy compared to that of block variance features.
- Includes contribution in feature extraction and pattern recognition for an increase in the accuracy of the system.
- Explains contourlet transform as the best modality-specific feature extraction algorithms for fingerprint, face, and palmprint.

This book is for researchers, scholars, and students of Computer Science, Information Technology, Electronics and Electrical Engineering, Mechanical Engineering, and people working on biometric applications.

Multimodal Biometric Identification System

Case Study of Real-Time Implementation

Sampada Dhole and Vinayak Bairagi

CRC Press
Taylor & Francis Group
Boca Raton London New York

CRC Press is an imprint of the
Taylor & Francis Group, an **informa** business

A CHAPMAN & HALL BOOK

Designed cover image: ShutterStock

First edition published 2025
by CRC Press
2385 NW Executive Center Drive, Suite 320, Boca Raton FL 33431

and by CRC Press
4 Park Square, Milton Park, Abingdon, Oxon, OX14 4RN

CRC Press is an imprint of Taylor & Francis Group, LLC

Library of Congress Cataloging-in-Publication Data

Names: Dhole, Sampada, author. | Bairagi, Vinayak, author.
Title: Multimodal biometric identification system: case study of real-time implementation / Sampada Dhole and Vinayak Bairagi.
Description: First edition. | Boca Raton: C&H/CRC Press, 2025. | Includes bibliographical references and index. | Summary: -- Provided by publisher.
Identifiers: LCCN 2024021491 (print) | LCCN 2024021492 (ebook) | ISBN 9781032660585 (hbk) | ISBN 9781032665979 (pbk) | ISBN 9781032665993 (ebk)
Subjects: LCSH: Biometric identification--Technological innovations--Case studies.
Classification: LCC TK7882.B56 D49 2025 (print) | LCC TK7882.B56 (ebook) | DDC 006.2/48--dc23/eng/20240907
LC record available at https://lccn.loc.gov/2024021491
LC ebook record available at https://lccn.loc.gov/2024021492

ISBN: 978-1-032-66058-5 (hbk)
ISBN: 978-1-032-66597-9 (pbk)
ISBN: 978-1-032-66599-3 (ebk)

DOI: 10.1201/9781032665993

Typeset in Times
by Deanta Global Publishing Services, Chennai, India

Contents

Preface

Most of the biometric systems presently being used are unimodal systems typically making use of a single biometric trait for identification purposes. There are several limitations of unimodal systems such as noise in sensed data, non-universality, intra-class variations, and inter-class similarities. Some of these can be addressed by designing multimodal systems that consolidate multiple sources of biometric information. Multimodal systems can improve the matching accuracy of a biometric system. They also address challenges such as non-universality, noise, susceptibility to spoof attacks, and large intra-class variations. Four levels of fusion are possible when integrating data from two or more biometric sources. These levels are sensor, feature, matching score, and decision levels. In multimodal biometrics, the identification system uses fusion at different levels using a combination of uncorrelated modalities (for example, face and palmprint images, two fingers of a person) and a combination of correlated modalities (for example, different fingerprint matchers, different face matchers). Still there are problems associated with fusion techniques and which fusion technique is the best is yet to be decided. Multimodal systems, if properly designed, are able to increase the matching accuracy of a recognition system as they consolidate the evidence from different biometrics, increase population coverage, and prevent spoofing attacks. This book mainly focuses on multimodal biometric identification and sensor-level, feature-level fusion.

The problems of unimodal systems can be efficiently dealt with the implementation of multimodal systems as described below

1) Limited population coverage is the issue of non-universality which is addressed in multimodal systems. If one biometric system acquires a poor multimodal biometric identification system using fusion of random subset of biometrics quality of data in the enrolling system, then other biometric data will acquire meaningful data and enrol the user.
2) Because of noisy data, the problems arise in unimodal systems. When the information acquired from one biometric trait is corrupted by noise, it is possible to use information acquired from the other biometric trait.
3) It is extremely difficult for an intruder to spoof multiple biometric traits simultaneously of a legitimately registered individual.

PURPOSE OF THE BOOK

a) Most of the publication proposes post-mapping fusion (score and decision level) of different modalities for various applications, but still pre-mapping fusion (sensor and feature level) in multimodal is still a largely unexplored area.
b) This book presented and explained a novel method of multimodal biometric fusion using random selection of biometrics, which covers a new method

of feature extraction, a new framework of sensor-level and feature-level fusion.

c) This book presented random selection of biometrics. Random selection of biometric traits would ensure that the system is interacting with a live user.

d) This book reveals that it is extremely difficult for an intruder to spoof fingerprint, face, palmprint, and hand geometry traits simultaneously of a legitimately registered individual as a subject unsuitable for one modality can use another modality (e.g., manual workers having cuts and bruises on their fingerprint).

e) This book focuses on fusion in biometric systems. It discusses the present level, the limitations, and proposed methods to improve performance.

BOOK COVERS THE PROBLEMS

The problems of unimodal systems can be efficiently dealt with the implementation of multimodal systems as described below

- Limited population coverage is the issue of non-universality which is addressed in multimodal systems. If one biometric system acquires a poor multimodal biometric identification system using fusion of random subset of biometrics quality of data in an enrolling system, then other biometric data will acquire meaningful data and enrol the user.
- Because of noisy data, problems arise in unimodal systems. When the information acquired from one biometric trait is corrupted by noise, it is possible to use information acquired from the other biometric trait.
- It is extremely difficult for an intruder to spoof multiple biometric traits simultaneously of a legitimately registered individual.

Author Biography

Sampada A. Dhole has completed his PhD in Electronics from Bharati Vidyapeeth (Deemed to be University) College of Engineering, India, in 2017 with specialisation in Image Processing and Biometrics. Her research interest includes the Image Processing and Multimodal. She has published more than 30 research papers including 7 Scopus indexed. She has filed 2 patents and 1 copyright in her technical field. She has worked as a Reviewer for many International and National Conferences. She is working as Assistant Professor in the Department of E&TC at Bharati Vidyapeeth's College of Engineering for Women, SPPU, Pune, India. She has 21 years of teaching experience. She is a member of the Technical Society ISTE, India.

Vinayak K. Bairagi has completed ME (Electronics) from Sinhgad COE, Pune, India, in 2007 (1st Rank in SPPU). Savitribai Phule Pune University has awarded him a PhD degree in Engineering. He has teaching experience of 16 years and research experience of 8 years. He has filed 12 patents and 5 copyrights in his technical field. He has published more than 60 papers, of which 26 papers are in international journals. He has authored/edited more than eight books/book chapters with multiple publishing concerns and he is a reviewer for nine scientific journals. He has received grants from DST SERB, UoP-BUCD, GYTI. He has received more than 14 awards, which include the National Level Young Engineer Award (2014), the ISTE National level Young Researcher Award (2015) for his excellence in the field of engineering, and IETE M N SAHA Memorial Award-2018. He is a member of INENG (UK), IETE (India), ISTE (India), and IEI & BMS (India). He had worked on Image Compression at the College of Engineering, Pune, under Pune University. His main research interests include Medical Imaging, Machine Learning, Computer-Aided Diagnosis, and Medical Signal Processing. Currently, he is associated with the AISSMS Institute of Information Technology, Pune, India, as Professor in Electronics and Telecommunication Engineering. He is a recognised PhD guide in Electronics Engineering of Savitribai Phule Pune University. Presently he is guiding seven PhD students.

1 Introduction

1.1 BIOMETRIC IDENTIFICATION SYSTEM

One of the primary concerns in the present day of information technology is the accessible and widespread availability of information that needs to be secure. Because the integrity and confidentiality of information are vital, it must be protected from unauthorised access. The term "security" refers to preventing (unauthorised) access to certain crucial information or valuables. For this reason, a variety of applications, including those for ATMs, driving licenses, passports, citizen cards, cell phones, and voter ID cards, require precise, automatic personal identification. Apart from identity, security holds equal significance. Passwords and PINs, which were utilised for identification and validation in previous decades, are unreliable due to the likelihood of fraud. Using biometric identification provides a solution to this kind of issue [1].

Biometrics is a combined word consisting of the Greek words "bios" meaning life and "metron" meaning measure. Biometric systems use behavioural or physical characteristics to identify individuals. The fingerprint, face, iris, retina, palmprint, speech pattern, signature, gait, and many other characteristics are most often utilised in biometric systems. The two main uses of biometric systems are authentication and identification. A biometric system has various benefits over conventional techniques. Biometric characteristics are not lost or forgotten, unlike tokens and passwords. Biometric characteristics are difficult to duplicate, share, transfer, or steal. A biometric system functions as a pattern recognition system by collecting biometric data from each user, extracting a feature set from the collected data, and comparing the feature set with a template that has been stored in the database.

The enrolment and identification modules make up the two modules of the biometric identification system. Enrolling each person's biometric data into the biometric system's database is the responsibility of the enrolment module. The user's input is compared to the template of every individual registered in the database in the identification module, and the biometric system outputs the identity of the person whose template corresponds most closely to the user's input.

Each biometric system is made up of four primary modules, which are outlined below.

1. Sensor module

A sensor module records a person's biometric information. A fingerprint sensor that records the ridge and valley patterns of a user's finger pictures is one example.

DOI: 10.1201/9781032665993-1

1

2. Feature extraction

After processing the collected biometric data, the feature extraction module isolates the most important information to create a fresh representation of the data known as a template. The database contains these features that were extracted. The feature extraction module, for instance, extracts the location and orientation of minute points in fingerprint images.

3. Matching

The matching module assesses the degree of similarity or dissimilarity between the extracted feature set and the templates kept in the system database.

4. Decision

The decision module uses the extracted characteristics' degree of similarity to the stored templates to either determine the user's identification or validate the identity that the user claims.

1.1.1 ENROLMENT MODULE

The following are the desirable properties of biometric characteristics for reliable recognition performance and good subject discrimination [1] (Figure 1.1).

- Universality
    Universality means every individual person should possess those biometric characteristics.
- Distinctiveness
    Any two individuals should differ enough from one another in terms of their biometric characteristics.
- Permanence

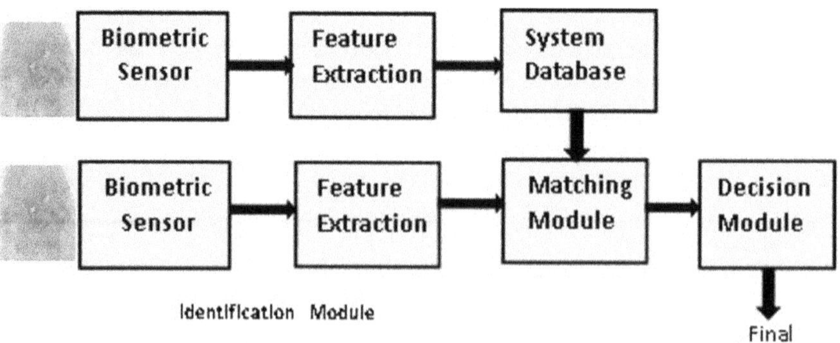

FIGURE 1.1   Typical biometric system.

Biometric features have to be sufficiently invariant over time (in terms of the matching criterion).

- Measurability

  The characteristics should be acquired easily. For further processing, the acquired raw data should be suitable.

- Performance

  Biometric characteristics should give high recognition accuracy.

- Acceptability

  Acceptability means people are willing to accept biometric characteristics.

- Spoof resistance

  Spoof resistance describes the level of difficulty of spoofing the biometric traits for physiological characteristics and behavioural characteristics (for example, fake fingers and mimicry).

1.2 CURRENT STATUS OF BIOMETRIC IDENTIFICATION SYSTEMS

Most of the biometric systems that are currently in use, known as unimodal systems, have a single biometric attribute to identify an individual. For applications involving identification and verification, these systems provide a dependable solution. However, it is crucial to take into account these systems' vulnerabilities and limitations.

It can be challenging to provide trustworthy authentication using unimodal biometric devices for a number of reasons. For example, the non-universality of relevant biometric features causes challenges with enrolment. The concept of non-universality suggests that there could be a portion of users who lack the biometric characteristic that is being obtained [1]. Biometric spoofing, or the ability to trick unimodal systems by using contact lenses with replicated patterns for iris detection, is a subject that is equally concerning [2]. Furthermore, the impact of environmental noise on the process of acquiring data can result in an inaccuracy that could render systems practically non-functional from the beginning [2]. For example, voice recognition-based biometrics deteriorate quickly in noisy surroundings. Similarly, differences in the facial image and lighting circumstances have a significant impact on the efficacy of face verification [3].

The following is a description of some of the challenges that the unimodal systems have been observed:

1. Noise in the data sensed

The detected information could be distorted or noisy. For instance, voice samples are altered by cold or fingerprint photos with scars. Noisy data is caused by malfunctioning or badly maintained sensors as well as unfavourable environmental factors.

2. Intra-class variations

When the user manipulates or modifies the properties of the sensor during authentication (e.g., the physical sensor replaces the optical sensor in the fingerprint) or the user interacts with the sensor incorrectly (i.e., incorrect facial pose), there will be an intra-class variation that arises.

3. Inter-class similarities

Multiple users' feature spaces may have similarities between different classes. For instance, many individuals in a biometric system share comparable traits, such as father–son or twin features.

4. Non-universality

Biometric systems may not collect useful biometric information from certain user groups. In some cases, the biometric fingerprint system may remove minute details in an individual's fingerprint due to the poor quality of the ridge pattern.

5. Spoof attacks

An impostor can try to mimic the biometric characteristic. The act of spoofing entails altering one's biometric characteristics to evade detection. Compared to physical biometric features, behavioural biometric qualities are more easily manipulated.

So unimodal biometric systems have limitations and face the following challenges:

- An increase in noise in sensed data reduces the accuracy.
- Failure to enrol (FTE) errors are increased by non-universality.
- A lack of uniqueness or individuality raises the false accept rate (FAR).
- Inter-class similarity raises FAR.
- Intra-class variance raises the false reject rate (FRR).
- Vulnerability to spoof attacks, which are typical for voice and signature.

Multimodal biometric systems have the potential to mitigate certain issues associated with unimodal biometric systems [4]. Multimodal biometric systems are those that integrate cues from two or more biometric sources in order to identify an individual. Most of the drawbacks of unimodal systems can be avoided by using multimodal systems for identification or verification. By using multimodal systems as outlined below, the issues with unimodal systems can be effectively resolved.

- Limited population coverage is the issue of non-universality which is addressed in multimodal systems. If one biometric system acquires poor quality data in the enrolling system, then the other biometric data will acquire meaningful data and enrol the user.
- Unimodal systems face challenges due to noisy data. It is feasible to use data from the other biometric characteristic when the data obtained from the first biometric feature has been altered by noise.
- It is quite difficult for an attacker to spoof a properly registered person while posing several different biometric features at once.

Additionally, multimodal biometric systems have certain drawbacks. Compared to unimodal biometric systems, they are more costly and demand greater computing and storage power. Users of multimodal systems typically experience some discomfort as enrolment and verification take longer. Finally, if a correct technique is not followed for merging the evidence presented by the various modalities, the accuracy of the system may actually decrease in comparison to the unimodal system.

1.3 APPLICATIONS OF BIOMETRIC SYSTEMS

Biometric technologies are used in many different applications because they provide reliable identification and evidence [4]. Some of these are mentioned below.

- Internet access, computer network access, e-commerce, physical access control, ATM or credit card transactions, mobile phone, medical records management, personal digital assistant (PDA), distance education, etc., are all uses of biometric technology.
- National identity cards, driving licenses, passports, Social Security Numbers (SSNs), welfare spending codes, and border crossings all use biometric technology.

Forensic medicine applications use biometric systems for the identification of the dead, paternity tests, crime investigations, and criminal identification.

A fingerprint identification technology is utilised at Walt Disney World in Lake Buena Vista, Florida, to make sure that the same person uses the ticket every day. The right and left index fingers of a person's fingerprints are used to link their entry into the US with a visa.

The Indian government provides every citizen with an Aadhaar, a 12-digit unique identifying number. Using fingerprint, iris, and facial biometrics, it is regarded as the largest national identification number initiative in the history of the globe. Face and fingerprint biometrics are utilised in India for property transactions and apartment registrations. Face, fingerprint, and signature biometrics are also utilised in the issuance of driving licenses and passports.

REFERENCES

1. Anil K. Jain, Arun Ross, and Salil Prabhakar, "An introduction to biometric recognition", *IEEE Transactions on Circuits and Systems for Video Technology, Special Issue on Image- and Video-Based Biometrics*, vol. 14, no. 1, Jan 2004, pp 4–20.
2. Ashish Mishra, "Multimodal biometrics it is: Need for future systems", *International Journal of Computer Applications (0975 – 8887)*, vol. 3, no. 4, 2010, pp. 28–33.
3. B.Y. Hiew, Andrew B.J. Teoh, and Y.H. Pang, "Touch-less fingerprint recognition system", *IEEE*, vol. 1–4, 1 Jul 2007, pp. 244–1300.
4. M. Golfarelli, D. Maio, and D. Maltoni, "On the error-reject trade-off in biometric verification systems", *IEEE Transactions on Pattern Analysis and Machine Intelligence*, vol. 19, no. 7, Jul 1997, pp. 786–796.

2 An Overview of Biometrics

2.1 BIOMETRICS

The word biometrics comes from the Greek word "bio" meaning "life" and metric means "measurement". Biometrics refers to the identification or identification of a person based on their physical and/or behavioural characteristics. Many authentication/identification-based biometric technologies have been developed based on the unique characteristics of the human body, ease of obtaining biometrics, public recognition, and hearing safety levels. This chapter provides an overview of the various biometric technologies in use/planning and their suitability for different tasks.

Motivation behind the invention of biometrics is the basic need for a person to be authenticated automatically and accurately. Authentication is the process of verifying that a user requesting a network resource is who he/she claims to be or not. Conventional authentication methods are based on two types:

- Something that you "have" – e.g., key, magnetic card, or smart card.
- Something that you "know" – e.g., PIN or password.

Biometric authentication method uses personal features, i.e.,

- Something you "are".

Simply put, biometric technology identifies people based on characteristics related to various physical, chemical, or behavioural characteristics. Biometrics is defined as a change in the purpose of reliably and quickly identifying and verifying a person's identity using unique biometric features and technology, thus solving crime.

Biometrics includes two types of characteristics as shown in Figure 2.1:

- Physiological: face, fingerprint, hand geometry, and iris recognition, and
- Behavioural: gait, odour, signature, and voice.

Biometric system uses the following human characteristics as identifiers:
Common:

- Fingerprint recognition
- Face recognition
- Speaker recognition

DOI: 10.1201/9781032665993-2

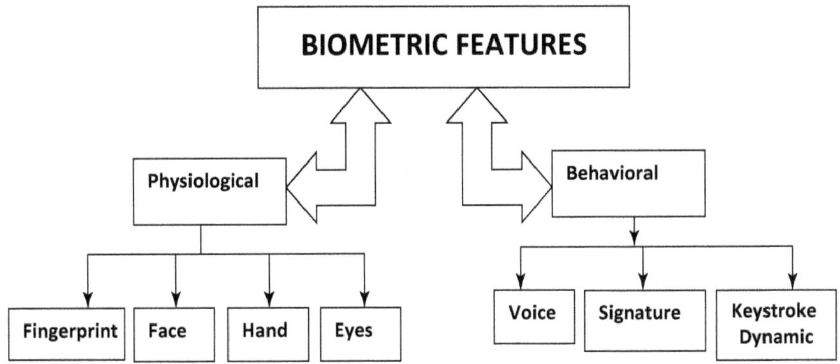

FIGURE 2.1 Types of biometric traits.

- Iris recognition
- Hand geometry
- Signature verification

Others:

- DNA
- Retina recognition
- Thermo grams
- Ear recognition
- Skin reflection
- Lip motion
- Body odour
- Brain wave pattern
- Footprint and foot dynamics

2.1.1 ADVANTAGES OF BIOMETRICS

The purpose of biometric technology and any device, system, or characteristic is to collect data and compare it for identification purposes. The advantages are:

- Universal among all people and specific enough to distinguish one person from another
- Permanent, may change over time
- Recordable (with or without consent)
- Measurable for future comparisons
- Anti-counterfeiting evidence (face, fingerprint)
- Real, enhanced system security, and fast authentication

2.1.2 DISADVANTAGES OF BIOMETRICS

Like other systems, biometric systems have advantages as well as disadvantages. Some of the disadvantages are:

- Injury, etc., due to physical disability causes changes in the body
- Expensive
- Physical characteristics cannot be changed
- Unstable assessment such as retina scan
- Software issues such as power issues
- No remote access and poor security

2.1.3 TYPES OF BIOMETRICS

Behavioural biometrics is the study of physical activity and is divided into:

- **Signature recognition**: The biometric signature recognition system identifies the identity and authorisation of authorised users.
- **Voice recognition**: Voice is a behavioural and physiological characteristic that depends on the structure of the throat and mouth and helps distinguish speakers as per the speech.
- **Keystroke:** Recognition based on keyboard strokes and sounds.

Physiological biometric identification is based on physical features such as ears, eyes, iris, and fingerprint and is divided into:

- **Ear authentication**: It relies on sound waves to determine the direction of the ear canal of a person.
- **Eye vein recognition**: Biometric identification uses the pattern of veins on the sclera (white layer of the eye) muscles.
- **Facial recognition**: 3D model of facial recognition.
- **Finger vein recognition**: Blood vessel pattern recognition under the skin of fingers.
- **Fingerprint recognition**: Fingerprint identification compares fingerprints with existing data, as no two people can have the same fingerprint.
- **Footprint and foot dynamics**: Footprint is a unique physiological character with distinctive properties useful in personal identification.
- **Gait recognition**: Gait refers to the cyclic and coordinated movement in human locomotion, and a person can be identified by observing his gait as a distinguishable feature.

2.2 FINGERPRINT

Fingerprint is defined as the characteristic pattern of ridges and valleys on the fingertip. Identification is based on local ridge characteristics.

The differentiation of the fingerprint characteristics or abnormal points on ridges is mentioned below:

1) Ridge ending: The point where a ridge ends abruptly.
2) Ridge bifurcation: The point where a ridge forks or diverges into branch ridges.

Fingerprint-matching techniques can be placed into two categories.

1. Minutiae-based technique
2. Correlation-based technique

2.2.1 MINUTIAE-BASED TECHNIQUE

This technique first finds minutiae points and then maps their relative placement on the finger. However, there are some difficulties when using this approach. It is difficult to extract minutiae points accurately when the fingerprint is of low quality.

2.2.2 CORRELATION-BASED TECHNIQUE

Correlation-based technique requires a precise location of the registration point and is affected by image translation and rotation.

The present work concentrates on the minutiae-based technique.

2.2.3 ADVANTAGES AND DISADVANTAGES OF FINGERPRINT BIOMETRICS

2.2.3.1 Advantages

- These systems usually are simple to use and install.
- Everyone's fingerprint is unique. Even a pair of identical twins has different fingerprints. This feature is used as a base for objectivity in personal identification.
- It requires inexpensive equipment which usually has a low power intake.
- A fingerprint pattern has an individually distinctive composition and its characteristic remains the same with time.
- Fingerprints are largely universal. Only 2% of the human population cannot use fingerprints due to skin damage or hereditary factors.
- Fingerprints are the most preferred biometric.
- Biometric fingerprint scanner presents a method to record an identity point which is very hard to be fake, making the technology incredibly secure.
- It is easy to use along with the high verification process speed and accuracy.
- It is one of the most mature and proven techniques (long history).

2.2.3.2 Disadvantages

- Because a fingerprint scanner only scans one section of a person's finger, it may be susceptible to error.

- Many scanning systems could be cheated by employing artificial fingers or perhaps showing another person's finger.
- Sometimes, it may take many swipes of fingerprints to register.
- Fingerprints of people working in chemical sectors are often affected.
- Cuts and marks transform fingerprints which often have negative effects on performance.

2.2.4 APPLICATIONS OF FINGERPRINTING

- Banking security – ATM security, card transaction.
- Physical access control (e.g., Airport).
- National ID systems, passport control (INSPASS), voting.
- Prisoner, prison visitors, inmate control.
- Secure E-commerce (still under research).
- A person's biometric identity is on the basis of one-to-one matches or on other security measures like password, PIN, etc.
- Identification of criminals and missing children.
- Secure e-commerce (still under research).
- One-to-one matches or passwords, PIN codes, etc., are used for a person's biometric identification.

2.3 IRIS RECOGNITION

Iris recognition is an automated method of biometric identification by employing mathematical pattern recognition techniques on the images of the iris of an individual's eye or unique complex random patterns.

2.3.1 ADVANTAGES OF IRIS TECHNOLOGY

Advantages of the technology are mentioned below:

- Accuracy: Iris patterns are unique so it is very accurate.
- Perfect protection: It is the inner part of the eye.
- Stability: Permanence of the iris structure.
- Non-invasive: Iris pattern acquisition is relatively easy.
- Unique: The possibility of two iris patterns being the same is nearly impossible.
- Flexible: Iris recognition technology can be easily integrated into existing security systems or operate independently.
- Stability: The iris pattern is stable for the whole life.
- Reliability: The unique iris pattern is not easy to be stolen, lost, or leaked.

2.3.2 DISADVANTAGES OF IRIS TECHNOLOGY

Iris biometrics disadvantages are listed below:

- Iris scanners might be very easily fooled through a superior quality image of an iris or face instead of the real thing.
- The scanning devices are often hard to adjust and may annoy multiple people of various heights.
- The accuracy of scanning devices may be impacted by unusual lighting effects and illumination from reflective types of surfaces.
- Iris scanners tend to be more expensive in comparison with additional biometrics.
- Because the iris is a tiny organ to scan from a long distance, iris recognition becomes challenging to perform well at a distance larger than a few metres.
- Iris recognition is vulnerable to inadequate image quality.

2.3.3 Applications of Iris Recognition System

Iris recognition is used in many applications such as:

- Used as a unique identification for bank accounts, Aadhaar card, etc., to the individuals.
- Used as password for computer login, for the bank cash machine accounts, and national borders security.
- Driving licences and personal certificates.
- Internet security, control of access to privileged information.
- Premises access control (home, office, laboratory).
- Anti-terrorism (e.g., security screening at airports).
- Used in financial transactions (electronic commerce and banking).
- Iris used in credit card authentication.
- Iris used in automobile ignition and unlocking; anti-theft devices.

2.3.4 Real-Life Applications

- Iris scanning forms the basis for Aadhaar card.
- Border patrolling in UAE.
- Iris scanning permits the passport in countries like US and Canada.
- Iris scanners are employed to control access of the datacentres of Google.

2.4 RETINAL PATTERN BIOMETRICS

Dr Carleton Simon and Dr Isodore Goldstein during 1930–1935 conducted their first scientific study to describe how every individual possesses a unique retinal structure. Because of the differing distribution of blood vessels, they suggested the use of the retina to confirm the identity of an individual. Dr Paul Tower (1950) reported that even identical twins have a very distinct and unique set of retinas and DNA strands leading to a very stable biometric modality that hardly changes. The human retina is a thin tissue composed of neural cells located in the posterior portion of the eye.

The retina is a complex structure with a network of blood capillaries and vessels resulting in a unique feature for every person as identical twins do not have the same features. Retinal patterns are altered due to medical morbidities, otherwise it is the same from birth to death. The unique retinal patterns of the individual are used as biometric identifiers, and the blood vessels within the retina absorb light more readily than the surrounding tissue which forms the typical smallest template for the biometric technology. The principle for the retinal scan is carried out by casting an unperceived beam of low-energy infrared light (IR) into a person's eye and the beam traces a standardised path on the retina as retinal blood vessels are more absorbent than the rest of the parts of the eye, then this path is converted to computer mode.

2.4.1 ADVANTAGES OF RETINAL RECOGNITION

- Due to stability and hardly ever changes over the lifetime, an individual's retina is a reliable biometric technology.
- Small-sized images give quick results for the identity of the individual.
- More number of unique data points which make errorless identity.
- The retina is present within the eye and, thus, not affected by the external environment like fingerprints.

2.4.2 DISADVANTAGES OF RETINAL RECOGNITION

- People are not willing to place their eyes on receptacles that are directly exposed to infrared radiation.
- Retinal recognition requires a high level of cooperation and motivation from the end user to capture high-quality images. This results in a lower verification rate of 85% compared to 100%.
- Non-involvement of end users leads to many rejections as it takes many attempts and a long time to get the final results.

2.5 FACIAL RECOGNITION BIOMETRICS

The facial features (eyes, nose, lips, and chin) of humans have played a significant role in the recognition of individuals for a long time. A face image can be acquired using a normal camera. It is the most common biometric for identity authentication. The two main approaches that are used to perform face recognition are the holistic or global approach and the feature-based approach.

Reasons why facial biometrics are more effective than biometrics:

- No physical impact is required to represent the user.
- Accurate with high enrolment and verification rates.
- There is no need for an expert to interpret the results.
- The only biometric system that allows true authentication in a one-to-many environment.

2.5.1 CHALLENGES IN FACE RECOGNITION

- Posture
- Lighting
- Expression
- Occlusion
- Delay
- Personal factors: gender

2.5.2 ADVANTAGES OF BIOMETRIC FACIAL RECOGNITION

- It does not require any cooperation of the test subject to do any work.
- Systems set up in airports, multiplexes, and other open public areas can easily identify an individual among the massive crowd.
- This performs massive identification which usually other biometric systems can't perform.
- The systems don't require any direct contact with a person in order to verify his/her identity. This could be advantageous in clean environments, for monitoring or tracking, and in automation systems.
- Incident monitoring for security with photos which in turn taken by using a digicam, but there may be no such evidence with the fingerprint technology to track these incidents.

2.5.3 DISADVANTAGES OF BIOMETRIC FACE RECOGNITION

- Face recognition isn't perfect and faces challenges in performing under certain conditions.
- One obstacle is associated with the viewing position of the face.
- Face recognition doesn't work effectively in bad/weak lighting, sunglasses/sunshades, lengthy hair, or other objects partly covering the subject's face.
- Not very effective for low-resolution images.

2.5.4 APPLICATIONS

Facial recognition biometrics have replaced PIN numbers and other applications such as:

2.5.4.1 Government Uses

- Law enforcement
- Security/crime prevention
- Immigration: Fast through customs
- Voter verification

2.5.4.2 Business Use

- Home security: Alert homeowners to learn of intruders.

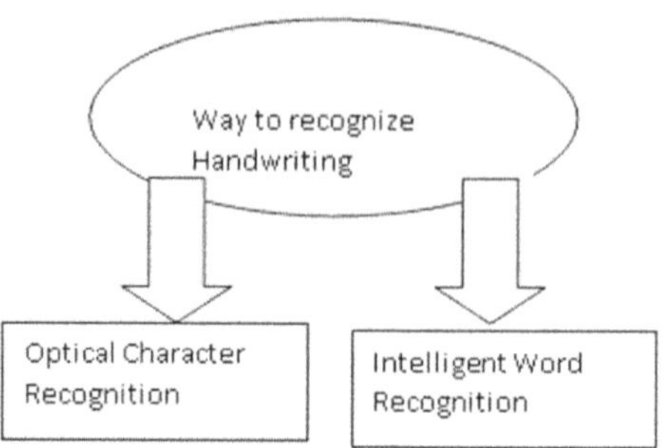

FIGURE 2.2   Ways to recognise handwriting.

- Banking using ATM: Quickly recognising users' faces through software.
- Body access.
- Facial recognition technology has now developed and become more afford-able, reliable, and accurate, thus finding application in many areas.

2.6 HANDWRITING

Handwriting detection is a computer technology that involves the process of receiving and interpreting comprehensible messages (e.g., written text) that can be typed from the source: documents, touchscreen, photos, etc. (Figure 2.2).

2.6.1 ADVANTAGES AND DISADVANTAGES OF HANDWRITING RECOGNITION

2.6.1.1 Advantages

Offline handwriting recognition has many advantages such as reading mailing addresses, bank check amounts, and forms. It forms the basis of the digital library and enables image data to be entered into computers through digitisation, image recovery, and recognition methods.

2.6.1.2 Disadvantages

The disadvantage of handwriting is that no information about the data or input can be obtained. First, an algorithm must be used to split the text into symbols or words and match these words in the data sequence.

2.7 VOICE BIOMETRIC

It is proved that the frequency, stress, and accent of speech differ from person to person. Voice biometry uses this concept to solve the problem of illegal users. This system has been implemented in the latest laptops as well.

2.7.1 ADVANTAGES

- Speech can be recommended as a natural input as it does not demand any training and it is considerably quicker as compared to some other input.
- Voice is usually a quite natural strategy to communicate, and in fact, it is not necessary for you to sit at a keyboard set or even work with a handheld remote control.
- This technique helps those people who have difficulty in using their hands.
- It does not require any training for users.
- It offers a big advantage to those who suffer from problems that may impact their writing capability, but they can use their voice to produce words/text on desktop computers or maybe other equipment.

2.7.2 DISADVANTAGES

- Even the most efficient voice recognition systems very often may make mistakes, when there is disturbance or some noise in the surroundings.
- Voice recognition systems work well only if the microphone is actually close to the end user. Much more far-away microphones are likely to boost the number of errors.
- May hacked with prerecorded voice messages.
- Possesses a primary amount of time for adjustment with each user's voice.
- Different persons might speak various languages.
- Several words sound very similar. Case: two, to, and too.

2.8 EAR RECOGNITION

Another biometric authentication technique is conducted based on the recognition of the unique shape and appearance of the human being's ear. Naturally, a person is born with a visual shape of his/her ears. However, the human ear is not subject to change while a person's growth and even ageing. It has a dependable stability which increases its level of security as a proposed method for the security identification/verification of individual

2.8.1 ADVANTAGES

It is a more comfortable/friendly method in terms of user contribution than iris and retina recognition.

2.8.2 DISADVANTAGES

As with other biometric recognition schemes, ear-based biometric authentication has its own drawbacks. This method has not achieved a remarkable level of security yet. One of the disadvantages of ear recognition is the simple distinguished features of the ear that cannot provide a strong establishment of an individual's identity.

2.9 SUMMARY

Biometric technology helps to identify a person based on their personal characteristics. Biometrics identifies people with fingerprints, facial features, retinas, etc. It is a tool that measures, compares, identifies, and defines according to characteristics. Personal characteristics such as hand geometry, iris colour, or personality make each person different and allow biometric screening, comparison, and identification through three types:

- Physiological
- Behaviour
- Physiology and behaviour

Authentication, identification, security, and accuracy have led to many applications of biometrics in technology:

- Office of administration
- Personnel information on citizenship and immigration
- Confidential and non-confidential information
- Identification crime evidence detection
- E-commerce security
- Reduce fraud and theft
- Enforcement of law

3 Motivation behind Multimodal Biometric Systems

3.1 INTRODUCTION

Even with significant recent progress, achieving trustworthy authentication with unimodal biometric systems remains very challenging. There are several causes for these. For example, enrolment issues arise from the fact that pertinent biometric features are not universal. The concept of non-universality suggests that there could be a portion of users that lack the biometric characteristic that is being obtained. Biometric spoofing, which implies that unimodal systems may be tricked by means of contact lenses with replicated patterns for iris recognition, is also concerning. Furthermore, the impact of ambient noise on the process of acquiring data can result in imprecision, which could render systems practically non-functional from the start. For example, voice recognition-based biometrics deteriorate quickly in noisy surroundings. Similarly, differences in the facial image and lighting circumstances have a significant impact on the efficacy of face verification. Multimodal biometrics can lessen some of the restrictions imposed by unimodal biometric systems. The main benefits of multimodal biometrics over unimodal biometrics are increased accuracy and resistance to spoofing. Multimodal biometrics makes it harder for an intruder to concurrently spoof the many biometric characteristics of a registered user by utilising complementing information. Moreover, the issue of non-universality is essentially resolved because a variety of features can guarantee a wide enough coverage population. Despite the fact that multimodal biometric systems require significantly more storage, processing time, and computational resources, they may be chosen over single modalities due to these benefits.

While the use of biometrics in identification has been researched for over 30 years, multimodal biometrics—which integrate data from multiple distinct traits—have attracted a lot of attention lately. Combining various traits has several benefits, including lowering the false accept rate (FAR) and false reject rate (FRR), strengthening systems against sensor failure, and preventing the system from reading input when there is scarring that obscures the fingerprint, among other situations. By assessing the error rates provided by FAR and FRR, many multimodal biometrics that have been proposed in the literature aim to improve recognition accuracy.

Multimodal biometric systems, which employ multiple biometric identities, can help overcome some of the drawbacks of unimodal biometric systems. Multimodal systems are more dependable than unimodal systems since they have numerous,

DOI: 10.1201/9781032665993-3

independent biometrics. Accuracy is higher since it is exceedingly difficult for imposters to spoof many biometric features at once [1, 2].

Four factors—the integration architecture, integration scenarios, level of fusion, and methodology for integrating the numerous cues—can be used to categorise multimodal biometric systems.

3.1.1 Advantages of Multimodal Systems over Unimodal Systems

Multimodal systems are those that combine biometric data and/or attributes from several sources in order to achieve the goal of identifying an individual citizen [1]. Most of the drawbacks of unimodal systems can be avoided by using multimodal systems for identification or verification. The reason for this is that distinct biometric sources typically compensate for the inherited issues associated with the other biometric traits [3]. As a result, compared to unimodal systems, multimodal systems have a number of advantages. Implementing multimodal systems, as outlined below, can effectively address the issues with unimodal systems:

- Limited population coverage, or non-universality, is addressed by multimodal systems. For instance, if a person's fingerprints are of poor quality and hinder him from enrolling in the system, the system can still enrol him by using other biometric traits like his voice, face, or palm print. These biometric traits assist the system in gathering useful biometric data.
- Multibiometric trait spoofing of a legitimate user is exceedingly challenging. If each subsystem determines the chance of the given trait being a spoof, it is possible to find out the probability of the user being an imposter by utilising an appropriate fusion technique. Additionally, a challenge-response system can be incorporated, asking the user to display a random selection of features (in a specific order) at the point of acquisition. This will ensure the interaction of the system with the live user.
- The issue caused by noisy data is efficiently addressed by multimodal systems. It is feasible to use data from the other biometric characteristic when the data obtained from the first biometric feature is tainted by noise. During the fusion process, certain systems also assess the quality of the biometric signals that were obtained as input. Determining the quality of obtained biometric data is a difficult task in and of itself. On the other hand, multimodal systems gain major benefits when implemented correctly.
- When data from specific biometric sources becomes inaccurate due to software bugs, purposeful user manipulation, or malfunctioning sensors, a multimodal system serves as a fault-tolerant system by continuing to function. In authentication systems that involve a large number of subjects, fault tolerance is typically desirable (for example, in border control applications).

The accuracy of biometric systems can be significantly increased by combining data from several sources. Improving matching accuracy depends on using appropriate fusion methodology and information sources. Multiple sources of information further expand the feature space, which raises the number of people who may be reliably discriminated against. As a result, an identity system's capacity—that is, the total number of people it can enroll—may be raised.

3.2   MULTIMODAL BIOMETRIC INTEGRATION ARCHITECTURE

A multimodal biometric system's architecture describes the sequence in which multiple features are obtained and processed.

- Serial architecture

As seen in Figure 3.1, modalities are processed sequentially in a serial or cascading design, with the results of one modality influencing the processing of succeeding modalities. Bank ATMs employ these types of architectures.

- Parallel architecture

Different modalities function independently in parallel architecture, and as shown in Figure 3.2, the outputs are integrated utilising a variety of fusion schemes. In military settings where a high level of security is necessary, these architectures are deployed.

3.3   MULTIMODAL BIOMETRIC INTEGRATION SCENARIOS

Figure 3.3 illustrates the five integration situations found in multimodal biometric systems [2].

1. Multiple instances

It combines various samples of two or more different instances of the same biometric characteristics captured by using one sensor: two impressions of a person's right index finger, for instance.

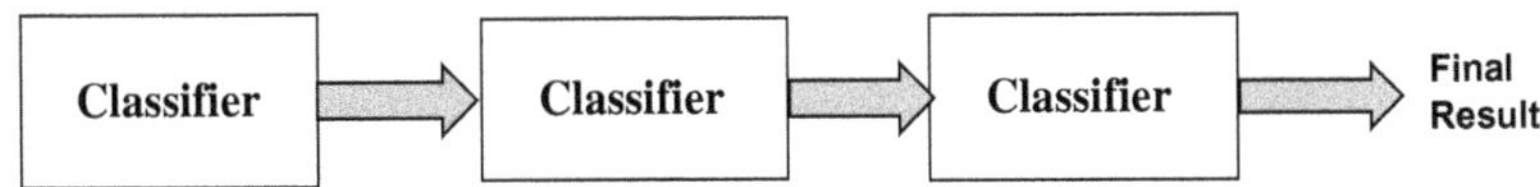

FIGURE 3.1   Serial architecture.

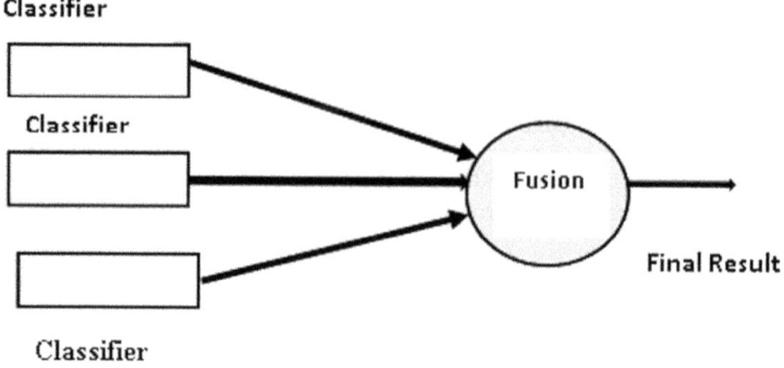

FIGURE 3.2　Parallel architecture.

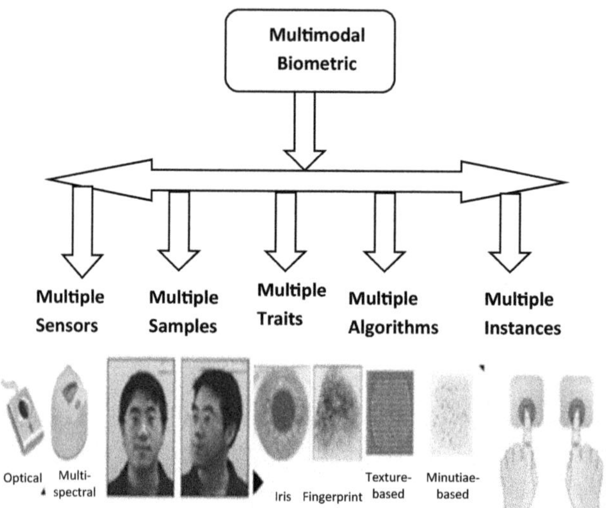

FIGURE 3.3　Integration scenarios of multibiometric fusion and sensors.

2. Multiple sensors

Here the combined output of various sensors is found. For instance, solid state and optical sensors are utilised to acquire fingerprint images, combining to produce a single image of a fingerprint.

3. Multialgorithm

For a single biometric, where the output of different algorithms are combined. For instance, a combination of texture-based and minutiae-based fingerprint matchers is used.

4. Multiple samples

Here various samples of the same biometric characteristics are combined. For instance, a single biometric feature can be obtained by combining the left and right iris.

5. Multiple biometric traits

Here various biometric traits are combined. For example, face and fingerprint traits are combined for the identification of a person.

3.4 MULTIMODAL BIOMETRIC FUSION LEVELS

As seen in Figure 3.4, the sensor and feature levels are seen as a pre-mapping fusion, but the matching score and the decision levels are regarded as a post-mapping fusion. In a pre-mapping fusion, the biometric data is incorporated before classification; in a post-mapping fusion, each biometric data is modelled independently, and all the biometric traits are fused following mapping into a decision space that matches. In a biometric system, the amount of data accessible for fusion decreases with each processing layer.

Four separate levels can be used for fusion: the sensor, feature, matching score, and decision levels. Because the input at each level is different, fusion at each level requires a separate computational task.

3.4.1 PRE-MAPPING FUSION

3.4.1.1 Sensor-Level Fusion

Multiple sensors can be used to combine a single biometric trait in order to extract features from captured photos. In this initial step of fusion, the raw data obtained from the same biometric characteristic utilising two or more sensors is joined (Figure 3.5).

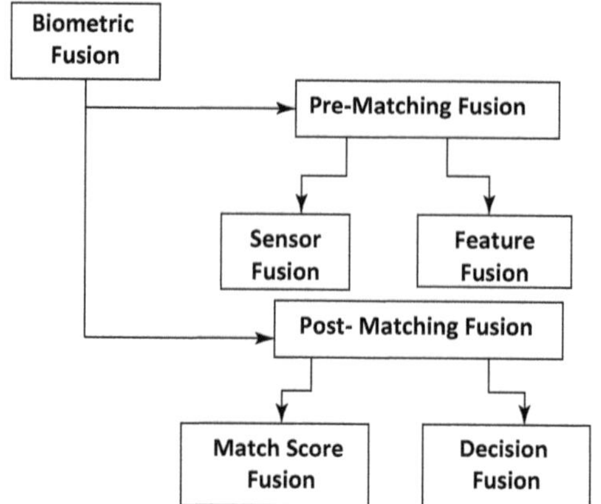

FIGURE 3.4   Biometric fusion levels.

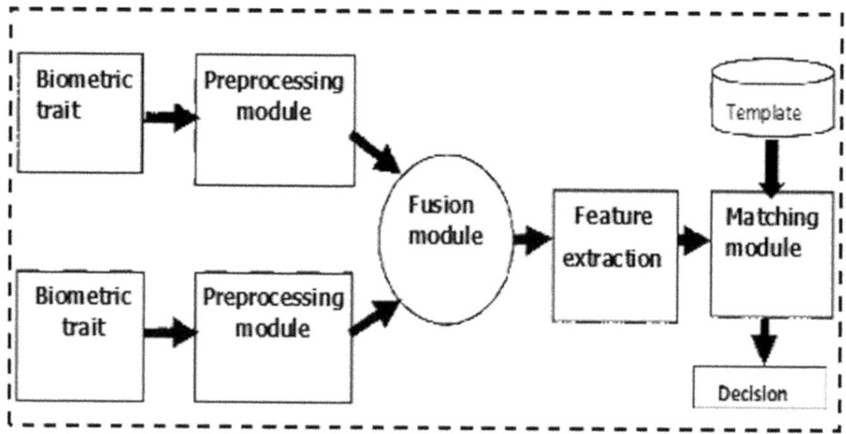

FIGURE 3.5   Fusion at sensor levels.

Current methods for sensor-level fusion combine numerous acquisitions of the same biometric characteristic from a single sensor. Combining two or more images to create a new one without distortion in the parts that overlap is known as image fusion or mosaicing. Sensor-level fusion is the process of combining several photographs taken with the same modality to create a composite image. A mosaicing approach for fingerprint images, as described by Ratha et al. [4] explains how several partial fingerprints are combined to create a rolling fingerprint that is obtained from the sensor surface. Given that it covers a bigger portion of the finger, this rolled fingerprint has more feature points. As a result, there is a greater matching between the rolled fingerprint template and the partial fingerprint picture (query impression) than there is between the two partial fingerprint images.

A multimodal biometric recognition system employing sensor-level integration of face and palm print images was proposed by Kishu et al. [5]. The Haar wavelet transform is used to break down images of faces and palm prints. Consideration is given to the fused image average of the wavelet coefficients. Ultimately, an inverse wavelet transform is performed to create a combined image of the palm print and the face. After that, features are extracted using the scale-invariant feature transform (SIFT) technique in order to reach a final accept/reject decision. Face and fingerprint samples are used in the Indian institute of Technology, Kharagpur (IITK) multimodal database to confirm the results.

Sensor-level fusion of visual and thermal face pictures is covered in several kinds of literature. For a face identification system, Singh et al. [6] described the sensor-level fusion of visible and thermal infrared pictures. According to the findings [6], face recognition ability in the infrared spectrum is substantially worse for images in the gallery when eyeglasses are present in the probe, and vice versa. To prevent facial occlusion due to eyeglasses, Singh [6] described the merging of information from both IR and visible spectra.

3.4.1.2 Feature-Level Fusion

Prior to feature vector extraction, signals from various biometrics are preprocessed. A composite feature vector is created by combining these feature vectors. By combining extracted features and using the proper feature normalisation, selection, and reduction algorithms, a single feature vector is produced (Figure 3.6).

Integrating feature sets that correspond to several modalities is known as feature-level fusion. Better recognition results are anticipated from integration at this level since the feature set contains more detailed information about the raw biometric data than either the match score or the decision. However, for the following reasons, fusion at this level is challenging to accomplish in practice.

(i) Different modalities' feature sets (such as a fingerprint's minute set or a face's eigen coefficients) might not work together.

(ii) It's possible that the relationships between the feature spaces of various biometric systems are unknown.

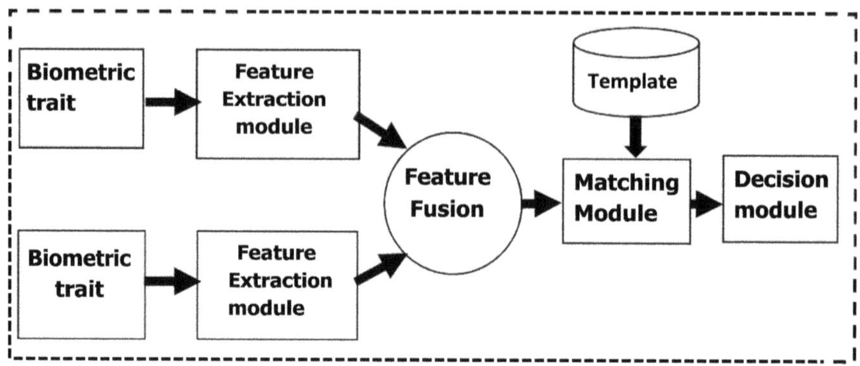

FIGURE 3.6   Fusion at feature levels.

(iii) The "curse of dimensionality" issue may arise by concatenating two feature vectors, which could produce a feature vector with extremely high
dimensionality.

Most feature fusion techniques use feature vector concatenation from several modalities, which affects the system's performance. A feature selection approach must be
employed in order to choose the best features because certain features may include
redundant information. When fusion is carried out on less discriminative data, performance may suffer in some situations because a fused feature vector may contain
noise from the data. A key component of achieving optimal system performance is
the feature vector.

Most feature fusion techniques use feature vector concatenation from several
modalities, which affects the system's performance. A feature selection approach
must be employed in order to choose the best features because certain features may
include redundant information. When fusion is carried out on less discriminative
data, performance may suffer in some situations because a fused feature vector may
contain noise from the data. A key component of achieving optimal system performance is the feature vector.

Fan Yang et al. [20] developed a novel mixed-mode biometric identification system using a fusion of fingerprint, hand geometry, and palmprint at feature level and
score-level fusion. DWT is used to extract features from fingerprint and palmprint
and extracted features are combined at the feature level to form joint feature vector.
For match score-level fusion, JFV and hand geometry feature scores are combined.
The performance results show the possibility of combinations.

Adams Kong et al. [21] improved the accuracy of palmprint identification system
using feature-level fusion. For feature extraction, Gabor filter is used with various
orientations to extract phase information on palmprint images. To produce single
feature called Fusion code, the output of all Gabor filter are merged. For matching of two Fusion code, hamming distance is used. For the final decision, dynamic
threshold is used.

Arun Kumar et al. [22] proposed recognition of palmprint and face-based multimodal using PCO-dependent feature level fusion is used. Evaluation is done with threshold value and fusion. If the threshold value increases, the FAR value also increases. The system shows 92% accuracy with FAR of 0.92% and FRR of 0.56%.

Praveen Kumar Nayak et al. [8] proposed a multimodal biometric system with face and fingerprint recognition using PCA and multilayer perception. System accuracy is 94% with FAR of 0.2% and FRR of 0.8%.

Krishneswari et al. [23] presented the fusion of fingerprint and palmprint at the feature level using wavelet-based image fusion techniques. Feature fusion based on a wavelet transform has several advantages in that the multiscale approach is able to manage image resolution where the image information is preserved in different kind of wavelet decomposition. For normalisation of features min–min approximation is used. Results show that system accuracy obtained is 98%, FAR of 1.02%, and FRR of 0.9%.

Rattani et al. [5] investigated feature-level fusion using face and fingerprint biometric modalities to improve the performance of the recognition system. They compare the performance of the system using sensor-level and feature-level fusion. Results show that feature-level fusion outperforms the score-level fusion by 0.67%. The system accuracy obtained is 98%.

Gayatri Bokade et al. [24] presented a multimodal biometric recognition system using face and palmprint at feature-level fusion. Simple fusion rules are applied to get combined image. As richer information is present in the feature set, fusion at feature level provides more accurate results as compared to other fusion level. The experimental results show GAR value of single modal palmprint is 81.48% and for face images 88.88%. After fusion of palmprint and face images at feature level GAR value obtained is 95% which shows substantial increase in accuracy.

Mitul Dhameliya et al. [25] developed and implemented multimodal biometric system using palmprint and fingerprint at feature-level fusion. For feature extraction Gabor filter with different orientations and scales are used. The results show 87% recognition rate with FAR 0.2% of and FRR of 1.1%.

3.4.2 POST-MAPPING FUSION

3.4.2.1 Score-Level Fusion

In this level, feature vectors are processed separately and an individual matching score is generated, and finally, these matching scores are combined to make the decision (Figure 3.7).

Amit Deshmukh et al. [7] propose a multimodal biometric recognition system using the fusion of face and fingerprint. Gabor wavelet is used for feature extraction of face images and local binary patterns (LBP) is used for fingerprint feature extraction. A weighted sum rule is used for the fusion of face and fingerprint images which provides better performance as compared to single biometric system.

Cheng Lu et al. [9] developed a multimodal biometric identification system using face and palmprint. For the extraction of features, two different techniques are used.

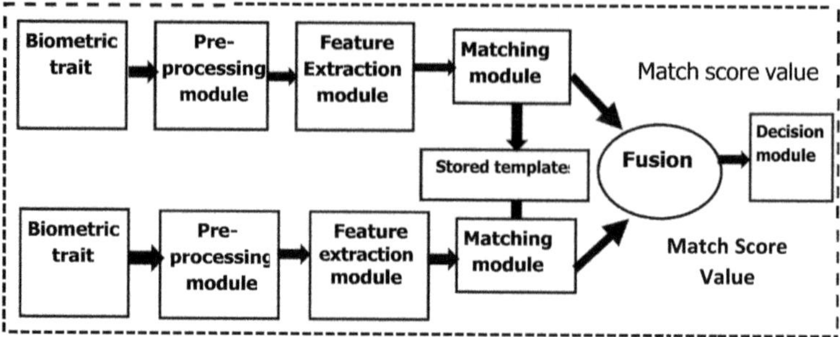

FIGURE 3.7   Fusion at matching score levels.

The first method is based on statistical properties and the second method is based on two-dimensional principal component analysis (PCA). The results show that the performance of a multimodal system is more than a unimodal system, and accuracy can reach up to 100% for the dataset considered in their research study.

Shefali Sharma et al. [10] presented a multimodal biometric personal verification system using hand shape and hand geometry. Only one acquisition device is required for shape and geometry features. The reference point at the wrist line is more stable than the centroid against the finger rotation and peaks and valleys determination. Both shape and geometry features are combined at the score-level fusion. For matching score-level fusion, the minimal distance rule is used.

Chaudhary et al. [11] proposed a multimodal biometric recognition system using palmprint, fingerprint, and face biometric modalities. To combine the features of palmprint, fingerprint, and face, a score-level fusion is used. Nageshkumar et al. [12] presented score-level fusion for palmprint and face image. The overall accuracy of the system is more than 97%, FAR 2.4%, and FRR 0.8%.

Rufeng Chu et al. [13] designed and implemented a multimodal biometric recognition system using a fusion of palmprint and face at score-level fusion. The results show that the recognition rate obtained is 98.85%, which validates that the accuracy of a multimodal system is more than its unimodal system. Cheng Lu et al. [9] designed and implemented a multimodal biometric recognition system using a fusion of palmprint and face at score-level fusion.

Muhammad et al. [14] explored a multimodal recognition system using the fusion of face and palmprint at the feature level. Features are extracted using a block-based discrete cosine transform, and to generate a fused feature vector, a matrix interleaved fusion is used. This nonlinear feature vector is used to train a Gaussian mixture model. The results show that the feature-level fusion gives better performance than score-level and decision-level fusions.

Navdeep Bajawa et al. [15] developed a multimodal biometric system using palmprint and fingerprint at the feature level. To develop a feature vector, a Gabor hidden Markov model is used. The proposed work was tested using the CASIA database.

Jain et al. [16] explored multimodal biometric recognition using the fusion of face, fingerprint, and hand geometry at score-level fusion. They analysed and presented the effect of different score normalisation techniques such as min–max, z-score, and tanh on the performance of a multimodal biometric system. The results confirm the improvement in performance after normalisation techniques were applied. The results show that the tanh normalisation technique is more robust than min–max and z-score. They demonstrated the sensitivity of both min–max and z-score methods to outliers. Min–max and z-score deliver good performance if the location and scale parameters of the matching scores (minimum and maximum values for min–max, or mean and standard deviation for z-score) of the individual modalities are known in advance.

3.4.2.2 Decision-Level Fusion

In this level of fusion, each modality is first pre-classified independently and then the final classification is based on the fusion of output at different modalities. Due to the availability of limited information, fusion at this level is considered to be rigid compared to the other fusion schemes.

Integration of information at the decision level can take place when each biometric matcher individually decides on the best match based on the input presented to it. Methods like majority voting [17], behaviour knowledge space [18], weighted voting based on the Dempster–Shafer theory of evidence [17], AND rule and OR rule, etc., can be used to arrive at the final decision (Figure 3.8).

Hong et al. [16] presented a multimodal biometric recognition system using the fusion of palmprint and face image at decision-level fusion. Recognition accuracy is improved by combining face and palmprint images at the decision level which achieves a recognition rate of 92%, FAR 1%, and FRR 1.8%.

Jain et al. developed a multimodal biometric recognition system using the fusion of face, fingerprint, and voice at the decision level. For integration, the product rule is used. Kumar et al. [19] presented the fusion of hand geometry and hand texture features using the sum rule at the decision level.

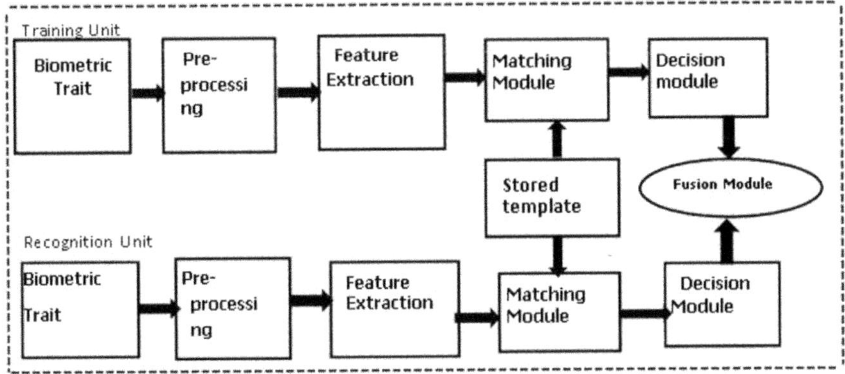

FIGURE 3.8 Fusion at decision levels.

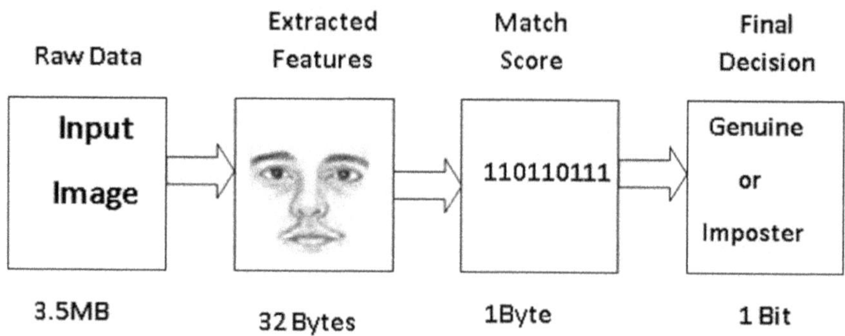

FIGURE 3.9 Example of how information available for fusion decreases at every level of a biometric system.

Fusion at the feature level will produce a new feature vector in a high-dimensional feature space. At the matching score level, the feature vector is the output of each expert, and thus it is reduced to a scalar value. At the decision level, the matching scores are compared with threshold values in order to derive the binary decision. Fusing at the matching score and decision level does not fully utilise important information in the feature vector because the integration is performed on scalar and binary values. The amount of information available to the fusion module decreases as we move from the sensor level to the decision level as shown in Figure 3.9. So integration at an early stage of processing is believed to be more effective.

REFERENCES

1. Ashish Mishra, "Multimodal biometrics it is: Need for future systems", *International Journal of Computer Applications (0975 – 8887)*, vol. 3, no. 4, 2010, pp. 28–33.
2. Robert Snelick, Umut Uludag, Alan Mink, Michael Indovina, and Anil Jain, "Large scale evaluation of multimodal biometric authentication using state-of-the-art systems", *IEEE Transactions on Pattern Analysis and Machine Intelligence*, vol. 27, no. 3, Mar 2005, pp. 450–455.
3. LinLin Shen a, Li Bai a, and Michael Fairhurst, "Gabor wavelets and general discriminant analysis for face identification and verification", *Elsevier- Image and Vision Computing*, vol. 25, 2007, pp. 553–563.
4. Ratha,N.K., Connell,J.H. and Bolle R.M., "Image Mosaicking For Rolled Fingerprint Construction", *In Proceedings of Fourteenth International Conference on Pattern Recognition(ICPR)*,vol.2, Brisbane, Australia,1998. pp.1651-1653.
5. A.Rattani, Kishu, Bicego, "Feature level fusion of faces and fingerprint biometrics", *1st IEEE International conference on biometrics, Theory, applications and systems,* 2007, pp1-6.
6. Singh, S., Gyaourova, A., Bebis, G. and Pavlidis, I., "Infrared and visible image fusion for face recognition", *SPIE Defense and Security Symposium*, 2004. pp.585-596.

7. Amit Deshmukh; Sheetal Pawar; Madhuri Joshi, "Feature level fusion of face and fingerprint modalities using Gabor filter bank", *IEEE International Conference on Signal Processing, Computing and Control (ISPCC),* 26-28 September 2013.

8. Praveen kumar nayak, Devesh Narayan, "Multimodal biometric face and fingerprint recognition using adaptive principal component analysis and multilayer perception", *International journal of research in computer and communication technology,* vol.2, no.6, 2013. pp 313-321.

9. Cheng Lu, Jisong Wang, Miao Qi, "Multimodal Biometric Identification Approach Based on Face and Palmprint", *Second International Symposium on Electronic Commerce and Security, IEEE computer society,*2009. pp 44-47.

10. Shefali Sharma, Shiv Ram Dubey, Satish Kumar Singh, Rajiv Saxena, Rajat Kumar Singh, "Identity verification using shape and geometry of human hands". Expert Systems with Applications, 2015. pp.821–832.

11. Sheetal Chaudhary, Rajender Nath, "A Multimodal Biometric Recognition System Based on Fusion of Palmprint, Fingerprint and Face", *International Conference on Advances in Recent Technology in Communication and Computing, IEEE computer society,* 2009. pp.596-600.

12. Nageshkumar, Mahesh.PK, Shanmuka swami M.N, "A Efficient Multimodal biometric fusion using palmprint and a face image", *International Journal of computer science,* Vol.2, Issue 3, 2009.

13. Rufeng Chu, Shengcai Liao, Yufei Han, Zhenan Sun, Stan Z. Li and Tieniu Tan" Fusion of Face and Palmprint for Personal Identification Based on Ordinal Features*", IEEE Conference on Computer Vision and Pattern Recognition,* 2007 (CVPR '07).

14. Muhammad Imran Ahmad, Wai Lok Woo, Satnam Dlay, "Non-stationary feature fusion of face and palmprint multimodal biometrics", Neurocomputing, 177, 2016. pp 49-61.

15. Navdeep Bajwa, Gaurav Kumar, "Multimodal Biometric system Based on Feature Level Fusion of Palmprint and Fingerprint Using Gabor Hidden Markov Model", *International Journal of Innovations in Engineering and Technology,* Vol.6, Issue 1, 2015. pp 87-93.

16. Lin Hong, Anil Jain, "Integrating faces and fingerprints for personal identification for personal identification", *IEEE Transactions on pattern analysis and machine intelligence,* Vol.20, No.12, 2008.

17. L. Lam and C. Y. Suen., Application of Majority Voting to Pattern Recognition: An Analysis of Its Behavior and Performance. IEEE Transactions on Systems, Man, and Cybernetics, Part A: Systems and Humans, 27(5), 1997. 553–568.

18. L. Lam and C. Y. Suen. Optimal Combination of Pattern Classifiers. Pattern Recognition Letters, Vol.16, 1995. pp 945–954.

19. A. Kumar, D.C.M. Wong, H.C. Shen, A.K. Jain, "Personal authentication using hand images", Pattern Recognition Letters vol.27 (13), 2006, pp.1478–1486.

20. Fan Yang, Baofeng Ma,"A New Mixed Mode Biometrics Information Fusion on Fingerprint, Hand-geometry and Palmprint", 4th international Conference on Image and Graphics, IEEE computer society,2007, pp.689-692.

21. Adams Konga,b, David Zhanga, Mohamed Kamelb "Palmprint identification using feature-level fusion", Pattern Recognition, vol.39, 2006, pp.478 – 487.

22. Arun kumar, Valarmathy, "Palmprint and face based multimodal recognition using PCO dependent feature level fusion", *Journal of theoretical and applied information technology,* 2013, Vol.57, no.3, pp 337-346.

23. Krishneswari K, Arumugam S, "Multimodal Biometrics using feature fusion", *Journal of computer science*, Vol.8, issue 3, 2012. pp 431-435.
24. Gayathri umakant bokade, Ashok sapkal, "Feature level fusion of palm and face for secure recognition", *International journal of computer and electrical engineering,* 2012, Vol.4, no.3, pp 157-160.
25. Mitul Dhameliya, Jitendra Chaudri, "A multimodal biometric recognition system based on fusion of palmprint and fingerprint", *International Journal of Engineering Trends and Technology (IJETT)* – Vol.4 Issue,5- May 2013, pp 1908-1911.

4 Performance Measurement Parameters for Biometric Systems

4.1 PERFORMANCE MEASUREMENT PARAMETERS

A person's fingerprint, palm print, or visage can all be used as biometric features for identification. The aforementioned attributes cannot match the supplied characteristic perfectly throughout the registration process. Variations in an individual's characteristics and the environment are the source of this. It may not be able to obtain the right characteristics during the registration process in the same environment due to changes in sensor location and different noise sources. The trait could change over time. The outcomes of all of this could mean that there isn't a perfect match for the person in the database. As it stands, just one outcome—a match that is as near to the original characteristic as possible—is preferred.

Thus, the simplest method that returns the closest match can be created for this. The issue arises when a person is not registered with the system, as it may nevertheless return the most similar match to that person. Naturally, it won't match as closely as that of a registered person, but since both registered and unregistered people are identified, the system is essentially rendered meaningless. This is avoided by using the threshold. Only matches that are over a particular threshold are considered valid; matches below that level are disregarded. The algorithm may produce inaccurate results even after applying a cutoff value to filter out the misleading acceptance of unregistered users.

Confusion matrix is given in Table 4.1.

The performance of a biometric system is measured in certain standard terms. These are

- False acceptance rate (FAR)
- False rejection rate (FRR)
- Equal error rate (EER)

i) **False acceptance rate (FAR)**

It is a ratio of number of people falsely accepted to the total number of people enrolled. It gives the probability of an impostor being accepted as a legitimate individual.

$$FAR = \frac{\text{Total False Acceptence}}{\text{Total False Attempts}}$$

DOI: 10.1201/9781032665993-4

TABLE 4.1

Confusion matrix

TP	FP
FN	TN

TP = True positive = correctly identified persons.
FP = False positive = incorrectly identified persons.
TN = True negative = correctly rejected persons.
FN = False negative = incorrectly rejected persons.

$$FAR = \frac{FP}{FP+TN} \tag{4.1}$$

ii) **False rejection rate (FRR)**

It is the ratio of the number of people falsely rejected to the total number of people enrolled. It gives the probability of a genuine individual being rejected as an impostor.

$$FRR = \frac{\text{Total False Rejection}}{\text{Total True Attempts}}$$

$$FRR = \frac{FN}{FN+TP} \tag{4.2}$$

iii) **Genuine acceptance rate (GAR)**

Biometric system accuracy is given by GAR. It is given by

$$GAR = 1 - FRR \tag{4.3}$$

GAR is important because it provides accuracy of the biometric system. We can directly compare two biometric systems by comparing GAR values of both systems. The system with the highest GAR value is considered to be a more accurate system.

iv) **Equal error rate (EER)**

On FAR vs FRR graph, the EER value is given by a point where both FAR and FRR curves coincide with each other. The false acceptance rate and the false rejection rate both depend on the threshold value, so to compare the two biometric systems the EER value is calculated. The system with less EER value is considered to be a more accurate system.

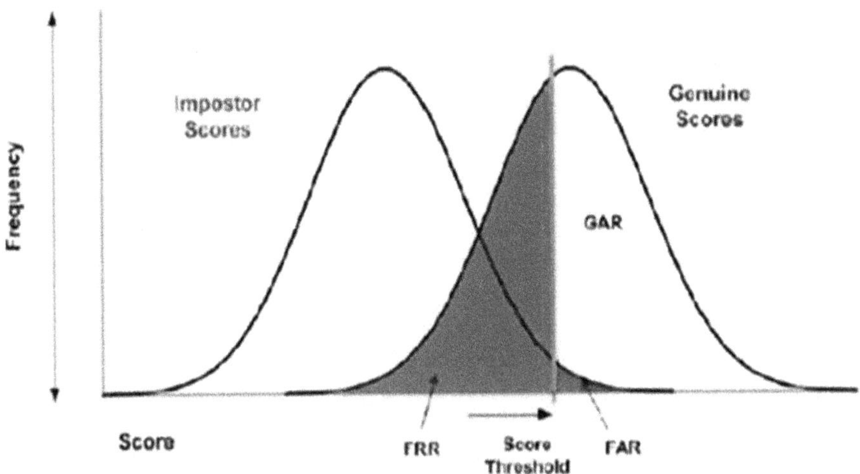

FIGURE 4.1 Receiver operating characteristics (ROC) curve [1].

Depending on the system's chosen threshold, a considerable amount of FAR and FRR may differ. In a similarity-based biometric matching system, a lower threshold will result in a lower FRR and a larger FAR. Similarly, when a similarity-based biometric has a greater threshold, the matching system's FRR will be higher and its FAR will be lower. A different measure called genuine accept rate (GAR) is occasionally used to determine a biometric system's accuracy. The GAR is calculated as the ratio of the number of individuals who are truly accepted (i.e., genuine people are accepted) to the total number of enrolled individuals for a predetermined threshold. Simply put another way, the GAR is calculated by deducting the number of wrongly rejected people out of the total number of legitimate people present.

Plotting the values of these performance criteria typically results in various graphs or curves that illustrate the biometric system's recognition accuracy. The receiver operating characteristic (ROC) plotting curve is the most widely used model for biometric verification. With varying thresholds, ROC curves display the FAR versus the matching FRR as showm in Figure 4.1. Equal error rate (EER), which is a point on a ROC curve, can also be used to indicate a biometric system's performance for equal values of FAR and FRR. Better performance is indicated by a lower EER value (Figure 4.1).

4.2 MATERIALS

The standard database (CASIA database) and the local database—which is our own captured database—are used for the experimental evaluations of the suggested approach.

4.2.1 Fingerprint Database

4.2.1.1 Standard Database

There are countless standard databases at one's disposal. Five hundred individuals' fingerprint scans are contained in the standard database utilised for this research project, the CASIA-Fingerprint database Version 5.0 (also known as CASIA-Fingerprint V5). The URU4000 fingerprint sensor was used to take the CASIA-Fingerprint V5 fingerprint images in a single session. There are waiters, employees, and graduate students among the CASIA-Fingerprint V5 participants. To create notable intra-class variances, the volunteers were asked to spin their fingers using varying pressures. Every fingerprint image is a 328 × 356 pixel resolution, 8-bit gray-level BMP file. Two hundred individuals were chosen at random for this study, and five samples from each person were taken into account.

4.2.1.2 Local Database

The other database for fingerprints was created by us. The database is created by taking fingerprints of 200 individuals and five fingerprints of each subject. The sensor U4500 fingerprint reader is used. U4500 fingerprint reader is a USB fingerprint reader designed and manufactured by Digital Persona with enterprise software applications and developer tools. This sensor is optical sensor with rotation invariant features. It gives a high-quality image with reliable performance over widest population of users.

4.2.2 Face Database

4.2.2.1 Standard Database

Several standard databases, such FERET and ORL, are accessible. Five hundred subjects' colour face images are included in the standard database utilised for this research project, the CASIA Face Image Database Version 5.0 (also known as CASIA-FaceV5). The Logitech USB camera is used to take a single session of face photographs for the CASIA-FaceV5. Waiters, employees, and graduate students make up the CASIA-FaceV5 volunteer pool. The resolution of all face images is 640 × 480 pixels, and they are all 16-bit colour BMP files. Variances in illumination, posture, expression, eyeglasses, and imaging distance are common intra-class variances.

4.2.2.2 Local Database

We also produced the other face database. A total of 200 distinct faces are used to form the database, with five faces belonging to each participant. The NIKON COOLPIX S8 digital camera is used to take face images. 7.1 megapixels and 3× optical and 4× digital zoom are features of the camera. The resolution of all face images is 640 × 480 pixels, and they are all 16-bit colour JPEG files.

4.2.3 Hand Database

For hand geometry and palm print recognition, the same database is used as palm print and is extracted from hand images.

4.2.3.1 Standard Database

All hand images of JPEG files with an 8-bit gray level are contained in the CASIA Multi-Spectral Palmprint Image Database. A CCD camera with different illuminations is utilised to take hand photographs. There are no pegs to limit hand or posture positions.

4.2.3.2 Local Database

The other database for hand was created by us. The database is created by taking hand of 200 individuals and five hand images of each individual. Hand images are captured using Digital Camera NIKON COOLPIXS8. Camera have 7.1 megapixels with 3× optical zoom + 4× digital zoom. All hand images are 16-bit colour JPEG files.

4.3 SUMMARY

While each biometric attribute works better on its own, there are situations in which it is ineffective. A multitude of challenges that occur with each unique biometric feature are addressed by the multimodal recognition system.

It is discovered via reading and evaluating the literature that:

- The choice of a biometric characteristic relies on its application.
- Biometric system performance is not optimum. The unimodal biometric system has limitations.
- Various biometric characters chosen must be compatible in fusion level used.

Multimodal biometric techniques can enhance the accuracy and performance of a biometric system.

- The most basic type of fusion with the least amount of information is decision-level fusion.
- Because scores have distinct distribution ranges, the score-level fusion requires the use of the normalisation approach.

The literature study reveals that, while being thought to produce better outcomes because fused feature vectors contain richer information, sensor-level and feature-level fusions have not been used as frequently. As a result, the study work that suggests a novel approach and examines feature fusion techniques at the sensor level as well as their usefulness in the process of recognition and verification is presented in the next chapter.

REFERENCE

1. Anil K. Jain, Arun Ross, and Salil Prabhakar, "An introduction to biometric recognition", *IEEE Transactions on Circuits and Systems for Video Technology, Special Issue on Image- and Video-Based Biometrics*, vol. 14, no. 1, Jan 2004, pp 4–20.

5 Unimodal Biometric Systems

5.1 UNIMODAL BIOMETRIC IDENTIFICATION SYSTEM

For digitisation, every biometric modality has to be scanned. To improve the accuracy of recognition, scanned images are pre-processed. Pre-processing consists of RGB to grey conversion, filtering, and binarisation. Segmentation is the key step after pre-processing, which extract region of interest (ROI). Pre-processing and segmentation are the essential steps before feature extraction. The effectiveness of the recognition and classification process depends on feature extraction, which should be able to quickly compute in order to extract more information while avoiding redundant data and decreasing noise. Figure 5.1 shows block diagram of Unimodal Biometric Identification.

To accomplish the recognition task and collect the biometric trait characteristics, a feature extraction stage is required before the pattern-matching stage. As a result of this feature extraction step, it would be ideal to have an accurate and reliable representation of the input signal while also keeping all of the crucial cues for identification. A reduced representation set of features, also known as a features vector, is created from input data that is too big for an algorithm to handle and is thought to be largely redundant (a lot of data, but not much information). Feature extraction is the process of converting the input data into a collection of features.

Feature extraction is the process of converting the input data into a collection of features. It is anticipated that the features set will extract pertinent information from the input data in order to accomplish the intended task utilising this smaller representation rather than the full-size input if the features extracted are correctly selected. This book suggests the discrete wavelet transform (DWT), Gabor, curvelet, and contourlet transform methods for feature extraction. and the outcomes are contrasted with the best practices for each strategy.

Two approaches are used for the experimentation in this book

1. The first approach uses the images of the standard database of fingerprints, palmprints, hands, and faces which are scanned by using various scanners and further processed.
2. In the second approach, the image is directly captured by using the camera and processed. The foregoing section describes the first and second approaches of experimentation.

DOI: 10.1201/9781032665993-5

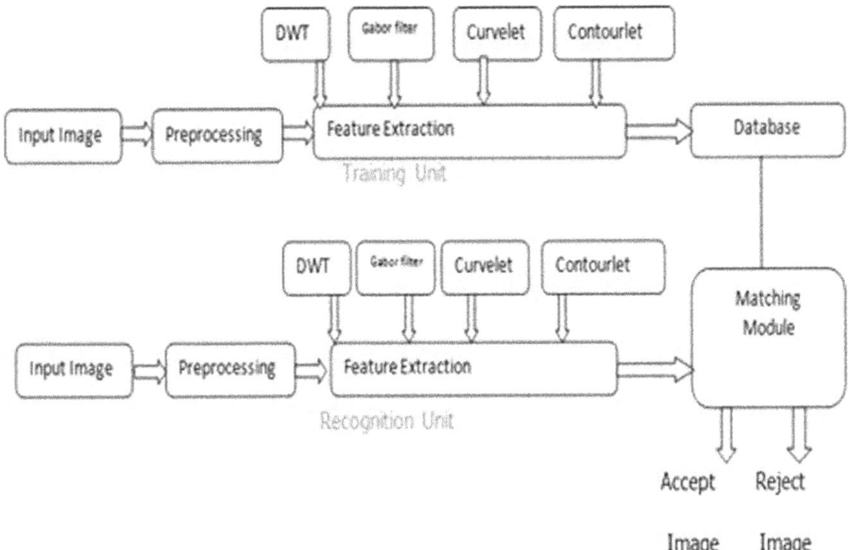

FIGURE 5.1 Unimodal biometric identification system.

5.1.1 DWT Feature Extraction System

One of the best methods for compressing images is the discrete wavelet transform (DWT). This transform works on the basis of hierarchically decomposing a signal into a multiresolution pyramid. At each resolution level, the signal is divided into a coarse approximation and some detailed information. In the following step, the approximation will be further broken down. The DWT is appealing because it offers excellent compaction features for numerous classes of natural images at a minimal implementation cost. At very low bit rates, each lossy compression algorithm exhibits its characteristic artefacts.

DWT reduces computing time significantly while providing enough information for the original image analysis. The row-by-row and column-by-column coefficient of an image is filtered successively with low-pass and high-pass filters to calculate the 2D DWT. The image's low-pass coefficient may undergo additional transformations following each transform pass. Recursive repetition of this process is possible depending on the situation. The approximation and detailed coefficient generation for the input $x(n)$ are displayed in Figure 5.2, where $g(n)$ and $h(n)$ represent the low-pass and high-pass filters' respective impulse responses for a single level of DWT. The image is divided into approximation and detail components by using DWT, which are represented by the sub-bands Low-Low (LL), High-Low (HL), Low-High (LH), and High-High (HH). Each sub-band's dimensions are half that of the main image. The variations in pictures or edges along horizontal and vertical directions are contained in the sub-bands HL and LH, respectively. The image's high-frequency information is contained in the HH sub-band. Figure 5.3 shows first level decomposition of DWT. Figure 5.4 shows Second level decomposition of DWT

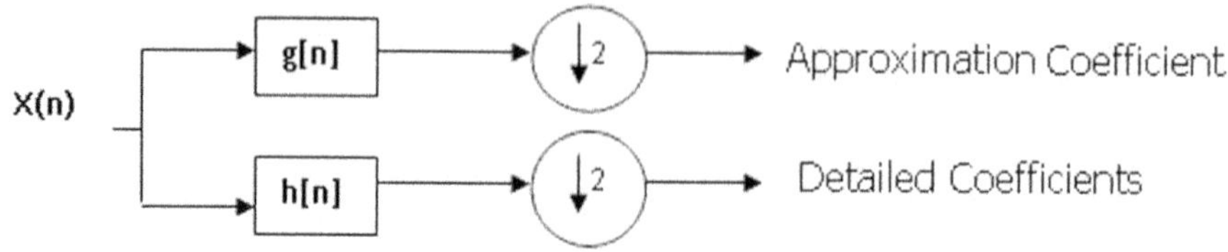

FIGURE 5.2   Block diagram of one level DWT.

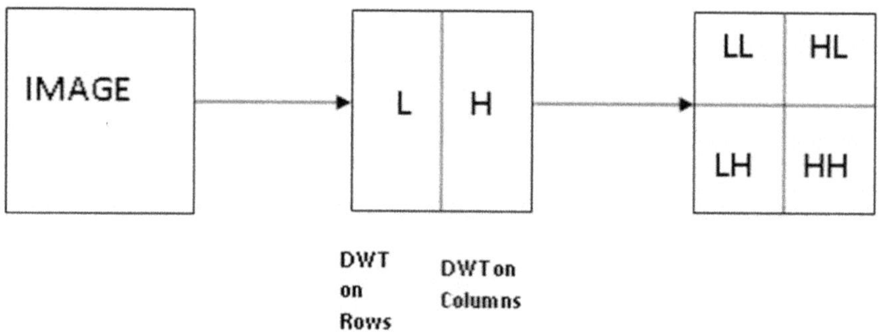

FIGURE 5.3   First-level decomposition of DWT.

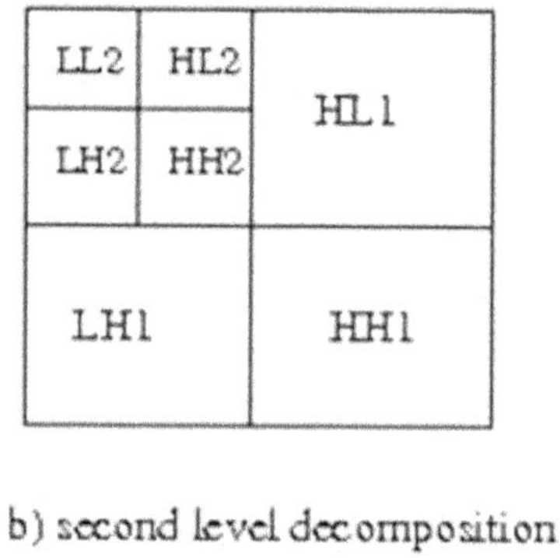

b) second level decomposition

FIGURE 5.4   Row–column computation of 2D DWT.

5.1.2   GABOR FEATURE EXTRACTION SYSTEM

It has been demonstrated that Gabor filters, one of the novel feature extraction methods employed in the literature, can extract the most information from small picture
regions and are invariant across scale, translation, rotation, and fluctuations in lighting [1, 2].

We automatically extract all the facial features using a bank of Gabor filters. The method is neither based on identifying certain and prominent facial features alone nor does it require any preparation of the input face image. In actuality, the system's foundation lies in identifying every feature point on the face that has high-energy regions. The primary benefit of this method is that even in cases where significant facial characteristics are obscured—for example, by spectacles, facial hair, or expressions—the algorithm should still be able to identify the face as accurately as possible using the data that were retrieved from other facial regions. Gabor filters are used to process facial photographs because of their biological relevance and a 2D Gabor filter in the spatial domain is a Gaussian kernel function modulated by a sinusoidal plane wave. It can be represented by

$$
G_{\theta,f}(x,y) = \exp\left\{\frac{-1}{2}\left[\frac{x'^2}{\partial_x^2} + \frac{y'^2}{\partial_y^2}\right]\right\}\cos(2\pi f x')
$$

$$
x' = x\sin\theta + y\cos\theta,
$$

$$
y' = x\cos\theta - y\sin\theta,
$$

where f = frequency of sinusoidal plane wave at an angle θ with the x-axis

∂_x and ∂_y = standard deviations of Gaussian envelope along the x- and y-axes, respectively.

Face picture features have been extracted using the Gabor filter bank, which consists of filters with varying orientations and frequencies. A Gabor filter bank with five frequencies and six orientations is typically utilised. Figure 5.5 displays the Gabor filter bank with five distinct scales and six distinct orientations. The Gabor filter bank has five frequencies ($m = 1, 2, \ldots, 5$) and six orientations ($n = 1, 2, \ldots, 6$).

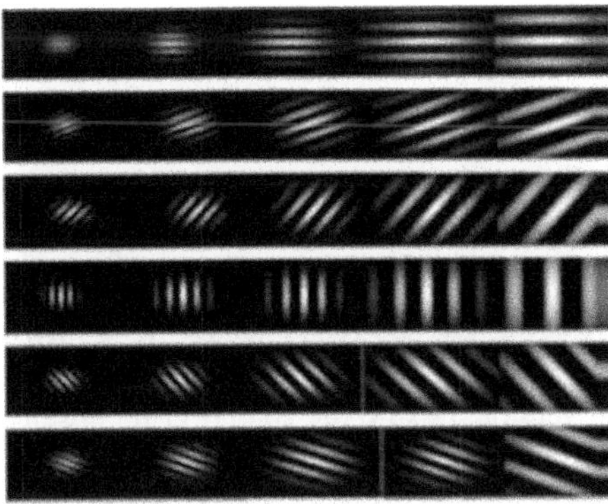

FIGURE 5.5 Real parts of the Gabor filters at five scales and six orientations.

5.1.2.1　Construction of Gabor Features

Once Gabor filters are created, the picture can be convolved with the filters to extract features (at various places) with varying frequencies.

5.1.2.2　SELECTING PEAKS/FEATURE VECTORS

From the image, we are able to extract every face feature. In order to extract the feature vectors/peaks from the image, an overlapping moving frame with a 5×5 size is created. Suppose these two requirements are met in the frame. In that case, the frame is scanned across to extract the peaks/feature vectors: local maxima are when any one pixel value in the frame is greater than or equal to every other pixel value in the frame; alternatively, all pixel values in the frame are greater than the average pixel value throughout the image.

5.1.3　CURVELET TRANSFORM

Wavelets are only effective at portraying point singularities since they don't make use of edge regularity or the geometric features of structures. Whereas curvelet transform captures all directions along wedges generated by curvelet decomposition, wavelet transform only allows the extraction of directional features in three dimensions: horizontal, vertical, and diagonal. Curvelets are represented by basis elements with elongated effective support—that is, length greater than width. Thus, the curvelet transform has extremely high anisotropy and directional sensitivity. There are four phases in the curvelet transform:

1. Sub-band decomposition.
2. Smooth partitioning.
3. Renormalisation.
4. Ridge let analysis.

Curvelet transforms of the second generation are quicker and less redundant than those of the first. With the removal of the ridgelet transforms in the new curvelet version, the transform's redundancy was decreased and its speed was significantly increased. Digital and continuous domain definitions for curvelet transforms are provided. Two distinct digital implementations of the second-generation curvelet transform exist: curvelets via wrapping and curvelets using USFFT (unequally spaced fast Fourier transform).

Comparing these new discrete curvelet transforms to their first-generation counterpart, they are faster, less redundant, and simpler. The identical digital colonisation is used in both digital versions, while the spatial grid chosen varies. Since curvelet via wrapping is currently the fastest curvelet transform available, we employed it in this instance.

A multiscale transform with a pyramid structure made up of several orientations at each scale is the wrapping-based curvelet transform. Curvelet can efficiently

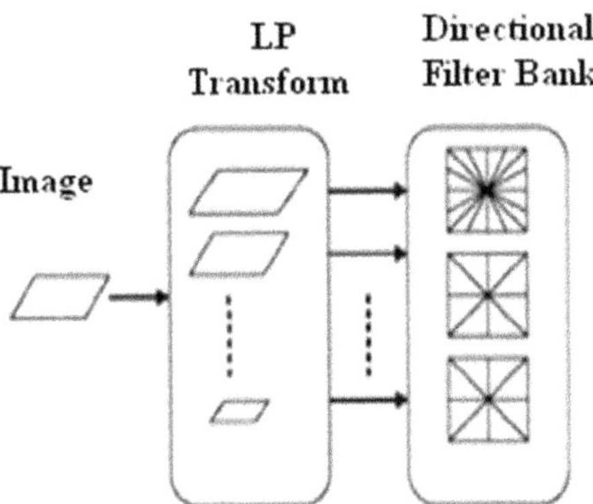

FIGURE 5.6   The contourlet transform framework [3].

capture curves in images by getting finer and smaller in the spatial domain and exhibiting more sensitivity to curved edges as the resolution level increases.

5.1.4 CONTOURLET TRANSFORM

The contourlet transform is an effectively directed multiresolution picture representation that Do and Vetterli recently presented. Because of its directionality and anisotropy, the contourlet transform performs better than the wavelet and curvelet transform in capturing the image's lines, edges, contours, and curves. The sub-band decomposition and the directed transform are the two steps that make up the contourlet transform. Point discontinuities are first captured using a Laplacian pyramid (LP), and then they are connected to a lineal structure using directional filter banks. It investigates the convergence of a discrete-domain building, which can be easily accessed by efficient methods, to a continuous domain expansion.

The contourlet-filter bank is a double-filter bank structure created by combining LP and directional filter bank (DFB). As seen in Figure 5.6, the band-pass images from LP are given to DFB in order to obtain directional information. As seen in Figure 5.7, the contourlet-filter bank divides the input picture into directional sub-bands at various scales.

5.2 FINGERPRINT AS A BIOMETRIC MODALITY

Fingerprint recognition system is simple, faster, and reliable. The fingerprint ridge pattern is constant throughout the life of a person. It does not change with person's age. Fingerprint is possessed by every human being, they are permanent and very distinct. They can change temporarily due to bruises and cuts but reappear after healing. Fingerprint is unique across individuals and even twins have different fingerprints.

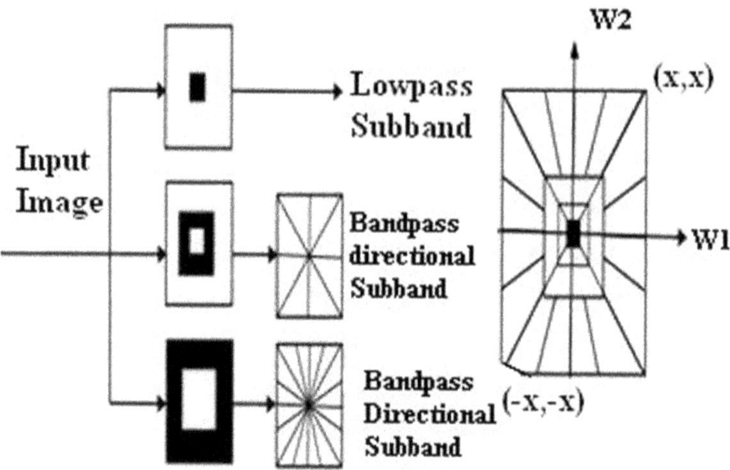

FIGURE 5.7   The contourlet-filter bank [3].

5.2.1 Techniques for Fingerprint Matching

- Minutiae-based matching
- Texture based

The minutiae-based approach is time consuming and requires very expensive processing techniques such as ridge filtering, thinning, segmentation, orientation estimation, normalisation, and binarisation, nevertheless, leads to incorrect minutiae. Previously, some form of post-processing had been applied to address this issue, although this removed both true and false minutiae. It is challenging to retrieve minutiae from really poor fingerprint photos. Though their distinctiveness is often lower, various elements of the fingerprint texture information may be recovered more reliably than minutiae. Thus, in the end, we can state that the robustness of the matching process is dependent upon the robustness of the fingerprint features that are extracted. Thus, a novel strategy that makes use of wavelet-based characteristics including DWT, Gabor, curvelet, and counterlet is provided. The experimental result from our own seized database and the CASIA database demonstrates that counterlet transform yields higher efficiency compared to other techniques.

5.2.2 Minutiae-Based Feature Extraction System

Two modules, enrolment and authentication, are present in the person identification system as shown in Figure 5.8. Feature vectors are created and stored in the database in the enrolment module. Stored database feature vectors are compared with extracted features in the authentication module. Identified or not identified results are displayed depending on matching scores. Figure 5.9 shows Minutiae Extraction process.

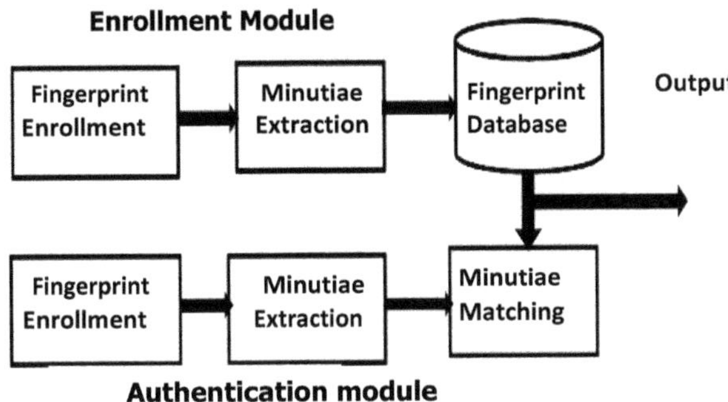

FIGURE 5.8   Fingerprint identification system.

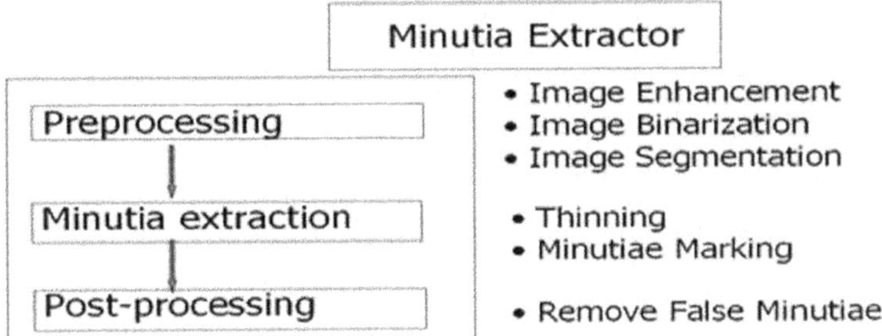

FIGURE 5.9   Minutiae extraction system.

The description of steps used for extraction of the actual minutiae points is as follows:

(i) **Pre-processing**

Pre-processing involves image enhancement, thresholding, image segmentation, and thinning. To show edge details, image enhancement is used. Histogram equalisation is used for enhancement, after that thresholding is utilised to convert a grey-level image to binary images. To get a single pixel width of an image, thinning is used.

(ii) **Minutiae marking**

Ridge endings and ridge bifurcation points are called as minutiae points as shown in Figure 5.10. Ridge bifurcation means the pixel that has two or more than two neighbours. Ridge ending means the pixel that has only one neighbour.

(iii) **False minutiae removal**

False minutiae points should be eliminated in order to improve recognition accuracy. Inadequate ink coverage leading to false ridge fractures and excessive inking causing ridge cross-connections are not entirely eliminated. Figure 5.11 shows results of Minutiae based fingerprint identification system

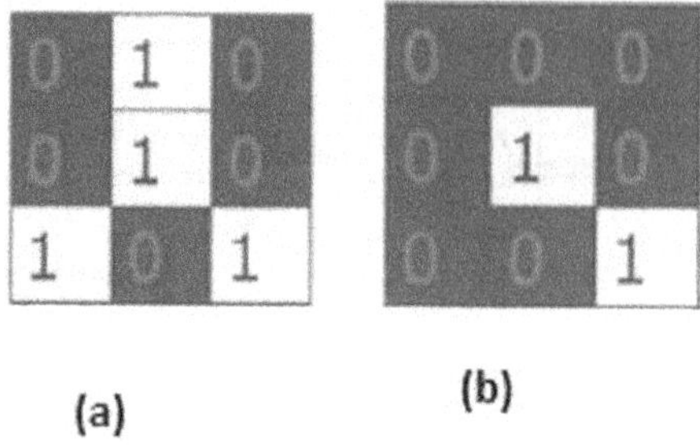

FIGURE 5.10   Minutiae points: (a) ridge bifurcation point and (b) ridge ending point.

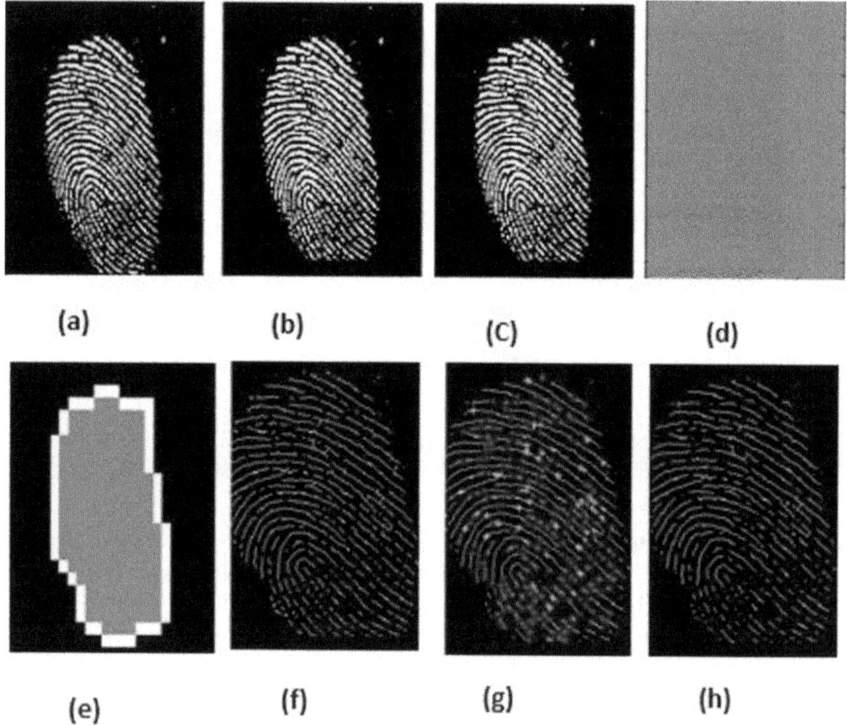

FIGURE 5.11   Results of minutiae-based fingerprint identification: (a) original image, (b) enhanced image after histogram equalisation, (c) binarised image, (d) directional ridge map, (e) ROI of the image, (f) thinned image, (g) minutiae points, and (h) real minutiae points.

5.2.3 TEXTURE-BASED FINGERPRINT RECOGNITION SYSTEM

The minutiae-based approach requires tedious and costly processing techniques such as segmentation, orientation estimation, ridge filtering, binarisation, noise removal, normalisation, and thinning, leading to inaccurate minutiae as well. To solve this issue, post-processing of some type can be applied; however, doing so also removes relevant details. Minutiae point extraction is highly challenging in cases where the fingerprint picture quality is inadequate. Though their distinctiveness is often lower, various elements of the fingerprint texture information may be recovered more reliably than minutiae. The quality of the fingerprint features that are extracted determines how strong the matching algorithm is. DWT, Gabor, curvelet, and contourlet transform are a few examples of wavelet-based features used in the newly suggested approach. Figure 5.12 shows Pre-processing steps for texture based fingerprint identification system

The segmentation, enhancement, and core point identification results are good in the texture-based recognition pre-processing step that is as shown below:

(i) **RGB to grey scale conversion**

RGB image converted to grey image using

$$\text{Intensity} = 0.2989 \times \text{red} + 0.5870 \times \text{green} + 0.1140 \times \text{blue} \qquad (5.1)$$

(ii) **Normalisation**

Normalisation is used to improve the quality of the image by removing noise and correcting image intensity deformations brought on by uneven finger pressure during image capture. Every pixel's intensity is adjusted during normalisation, resulting in a change in the mean and variance of the entire image to the set values.

An n-dimensional greyscale image is transformed using normalisation.

$$\text{IN} = (\text{I}-\text{Min}) \times [(\text{New Max}-\text{New Min})/(\text{Max}-\text{Min})] + \text{New Min} \qquad (5.2)$$

(iii) **Core point detection**

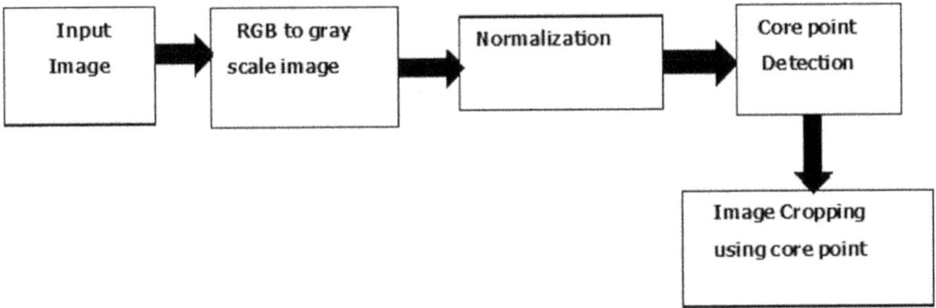

FIGURE 5.12 Pre-processing steps for texture-based fingerprint identification.

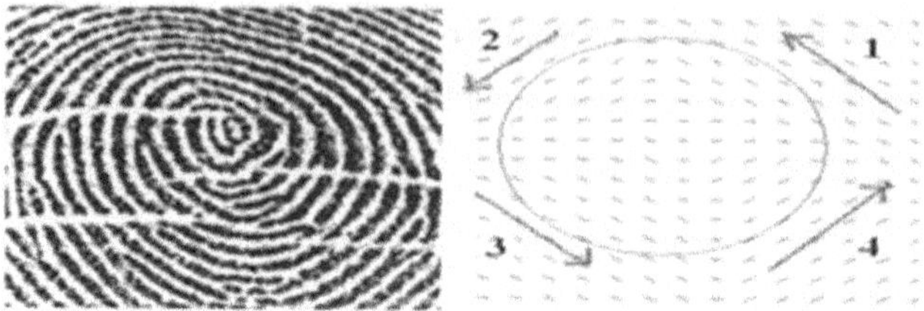

FIGURE 5.13   Mask of orientation field.

For determining areas with high curvature, core point is helpful. In order to locate well-formed arches, the orientation picture is scanned (row by row) within the core; an adjacent element that has six points indicates a well-formed arch. By assessing the orientation of the elements in nearby rows, one sextet is selected from among the valid sextets. A field mask technique is used to discover the orientation of the core point. Figure 5.13 displays an example of an orientation field in the core point region.

The range of angle (45–90 degrees) is present in regions 1, 3 and the angles (0–45 degrees) are present in regions 2, 4. Such pattern exists in a fingerprint are called as a core point.

(iv) Image cropping using core point detection

Cropping is extraction of a specific portion of the image and discards less useful information areas of the image. Figure 5.14 shows Result of Texture based Fingerprint Identification System

Table 5.1 shows comparison between texture-based fingerprint identification system for the local database. Results show that the contourlet transform equal error rate

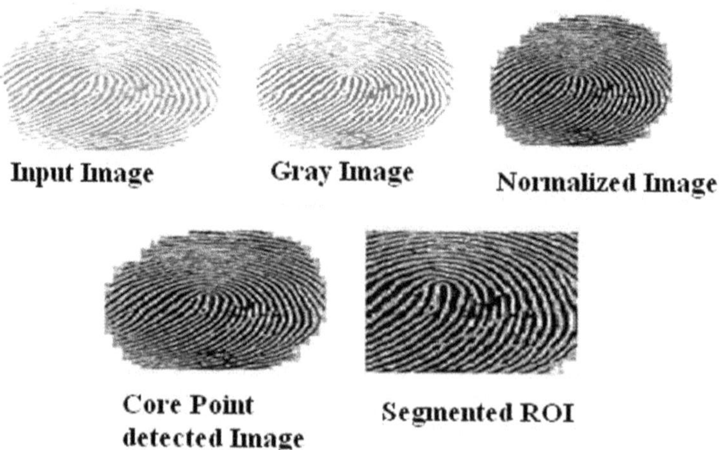

FIGURE 5.14   Results of texture-based fingerprint identification system: (a) original image, (b) RGB to grey image, (c) normalised image, (d) core point detected, and (e) cropped image.

TABLE 5.1

Different fingerprint identification techniques for the local database

Fingerprint methods	FAR	FRR	EER	GAR
DWT	0.984	0.1241	3.8	90.59
Gabor filter	0.8843	0.073	3.46	92.7
Curvelet	0.9714	0.0686	0.93	93.14
Contourlet transform	0.9397	0.0542	0.71	94.58

(EER) value is less as compared to other techniques, which means the performance of contourlet transform is better than DWT, Gabor filter, and curvelet transform.

Figure 5.15 shows the graph of the ROC curve (false acceptance rate (FAR) vs genuine acceptance rejection rate (GAR)). It describes comparison between texture-based fingerprint identification system for the local database. Results show that the contourlet transform GAR value is more as compare to other techniques, which

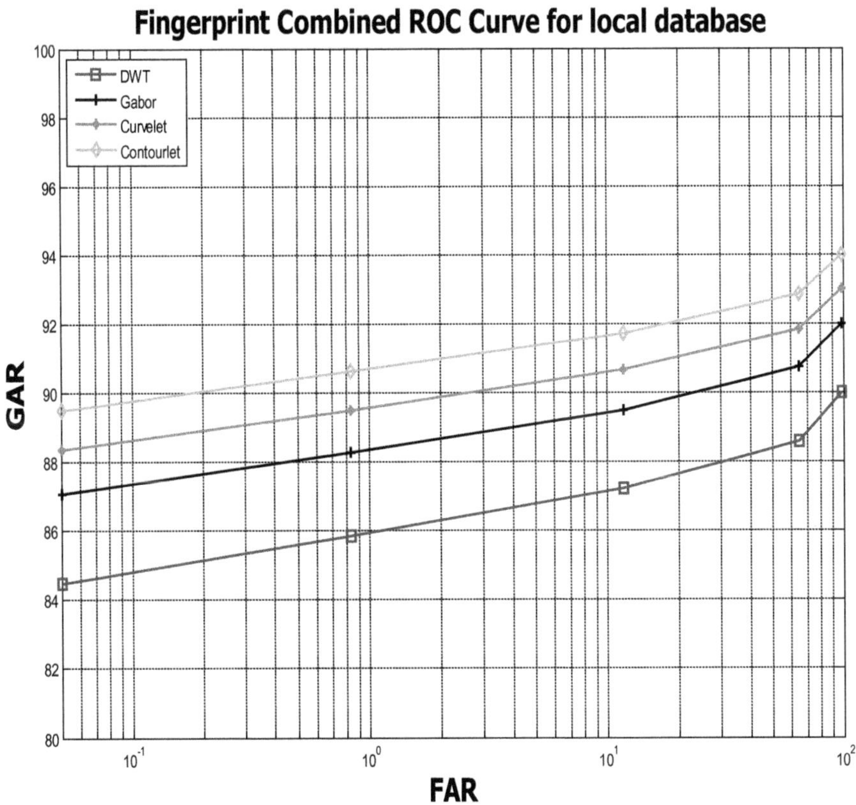

FIGURE 5.15   Unimodal fingerprint combined ROC for the local database.

TABLE 5.2

Different fingerprint identification techniques for the CASIA database

Fingerprint methods	FAR	FRR	EER	GAR
Minutiae based	1	0.13	0.97	87
DWT	0.8667	0.095	0.8505	90.5
Gabor filter	0.8926	0.0667	0.857	93.3226
Curvelet transform	0.92	0.0505	0.78	94.95
Contourlet transform	0.96	0.0495	0.72	95.0588

---

means the performance of contourlet transform is better than DWT, Gabor filter, and curvelet transform.

Table 5.2 shows a comparison between minutiae-based and texture-based fingerprint identification system for the CASIA database. Results shows that the contourlet transform EER value is less as compared to other techniques, which means the performance of contourlet transform is better than DWT, Gabor filter, and curvelet transform.

Figure 5.16 shows the graph of ROC curve (FAR vs GAR). It describes comparison between minutiae-based and texture-based fingerprint recognition systems for

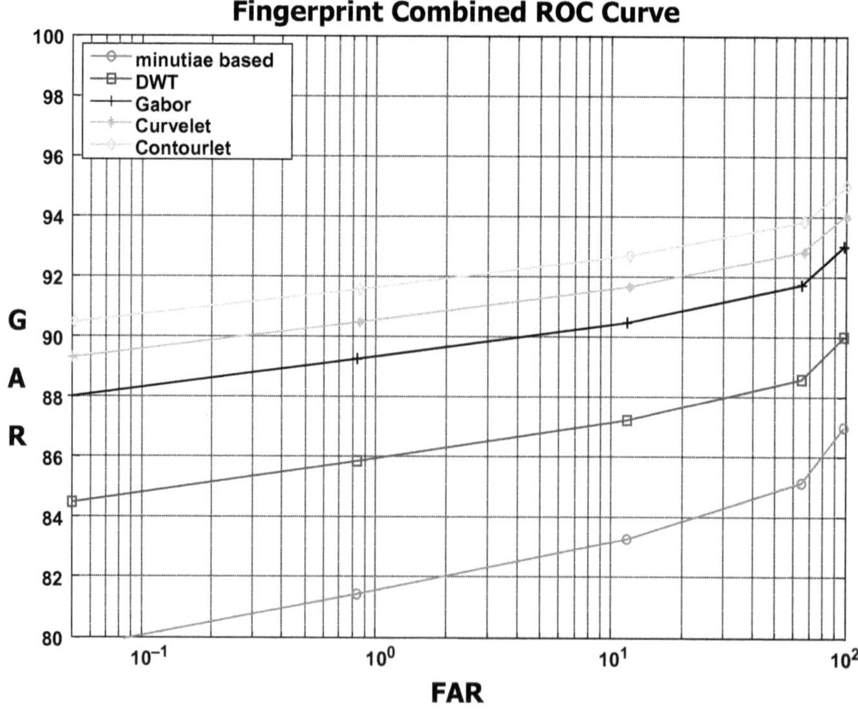

FIGURE 5.16 Unimodal fingerprint combined ROC for the CASIA database: (a) original RGB image and (b) grey image.

the CASIA database. Results show that the contourlet transform GAR value is more as compared to other techniques, which means the performance of contourlet transform is better than DWT, Gabor filter, and curvelet transform. The experimented results given in Tables 5.1 and 5.2 show that contourlet transform works better for fingerprint recognition systems.

5.3 FACE AS A BIOMETRIC MODALITY

Face recognition is best application in biometrics than fingerprint and iris recognition because it does not require physical contact like fingerprint. It is capable of capturing images at any moment from a distance, so it could be applied to any security application, such as debit card fraud, ATMs, and numerous others. Facial recognition welcomed by most users as it is non-intrusive, hands-free, and continuous. However, the following issues with facial recognition exist:

- The facial features varies with angle of orientation of the face.
- Face images change with age, make-up, presence or absence of spectacles, and even the expression on the face, such as furious, joyful, and sad.

About 75% of face authentication fails because of orientation angle; so, this research presented a wavelet-based face recognition technique that is robust to face picture orientation under identical illumination conditions.

5.3.1 TEXTURE-BASED FACE RECOGNITION SYSTEM

For face recognition, appearance-based features are considered. Here image is represented into long vector of grey levels. But it increased the dimensions of each face vector. To reduce dimensionality of the face image, principal component analysis (PCA) is used where only a few most significant Eigen vectors are considered to form a feature vector. But face images with different poses and expressions are non-linearly distributed, thus using linear space reduction technique cannot fully exploit information contained in face images so recognition accuracy is reduced. To overcome these problems, texture-based face recognition is presented.

(i) **Pre-processing**

In face identification algorithms to remove noise in images, pre-processing algorithms are used. Figure 5.17 shows after image RGB to Gray Conversion

(ii) **RGB to grey-scale conversion**

RGB image is converted to grey image using

$$\text{Intensity} = 0.2989 \times \text{red} + 0.5870 \times \text{green} + 0.1140 \times \text{blue} \tag{5.3}$$

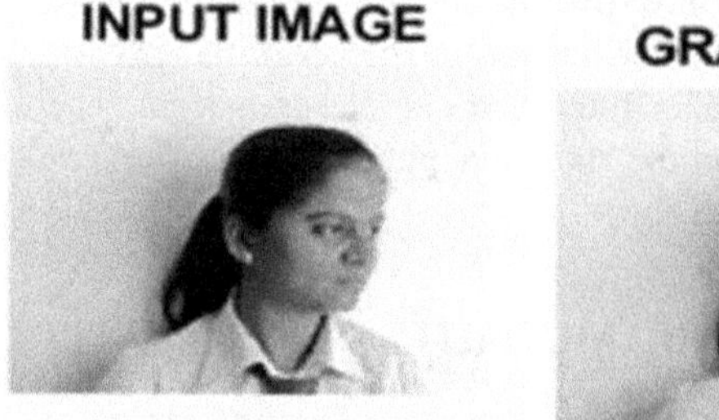

FIGURE 5.17   RGB to grey converted image.

(iii) Histogram Equalisation

Histogram equalisation is the process of modifying the intensity distribution on a histogram to increase an image's overall contrast. The most common intensity values are efficiently spread out via histogram equalisation, allowing greyscales to occupy all ranges from 0 to 255. Figure 5.18 shows Histogram of Face image before Equalization and after equalization

Histogram transformation is given by

$$S = T(r) \tag{5.4}$$

where $T(r)$ is in the interval range of $0 \leq r \leq 1$.

The likelihood that a level i pixel will appear in the image is given by

$$p_x(i) = p(x = 1) = \frac{n_i}{n}, 0 \leq i < L \tag{5.5}$$

where n_i be the number of occurrences of grey level i.

X = greyscale image.

L = No. of grey levels in the image; n = total number of pixels in the image.

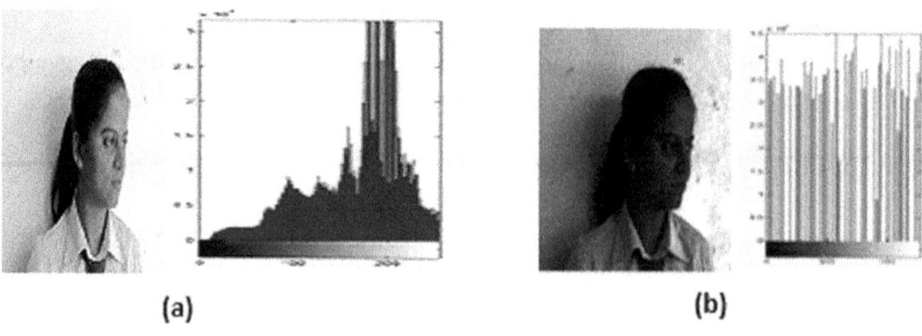

FIGURE 5.18   Histogram of the image before and after equilisation.

TABLE 5.3

Comparison of different face identification techniques for the local database

Face methods	FAR	FRR	EER	GAR
DWT	0.967	0.105	0.98	89.5
Gabor	0.8667	0.0953	0.9102	90.4762
Curvelet transform	0.902	0.0826	0.8165	91.74
Contourlet transform	0.8551	0.0758	0.78	92.42

The cumulative distribution function is given as

$$\mathrm{cdf}_x(i) = \sum_{j=0}^{i} p_x(j) \tag{5.6}$$

Table 5.3 shows comparison between texture-based face identification system for the local database. Results show that the contourlet transform EER value is less as compared to other techniques, which means the performance of contourlet transform is better than DWT, Gabor filter, and curvelet transform.

Figure 5.19 shows the graph of ROC curve (FAR vs GAR). It describes comparison between texture-based face identification system for the local database. Results show that the contourlet transform GAR value is more as compared to other techniques, which means the performance of contourlet transform is better than DWT, Gabor filter, and curvelet transform.

Table 5.4 shows a comparison between texture-based face recognition systems for the CASIA database. Results show that the contourlet transform EER value is less as compare to other techniques, which means the performance of contourlet transform is better than DWT, Gabor filter, and curvelet transform.

Figure 5.20 shows the graph of ROC curve (FAR vs GAR). It describes comparison between texture-based face recognition systems for the CASIA database. Results show that the contourlet transform GAR value is more as compare to other techniques, which means the performance of contourlet transform is better than DWT, Gabor filter, and curvelet transform.

5.4 HAND GEOMETRY AS A BIOMETRIC MODALITY

These days most researchers are attracted towards hand geometry-based personal identification systems due to great importance in many circumstances. The dimensions of the human hand, including its form, palm size, and finger lengths and widths, are the basis for hand geometry recognition systems. It is very easy to use, and environmental conditions like dry skin or dry weather have no detrimental effects on the accuracy of recognition. It is possible to obtain the hand photographs with a basic

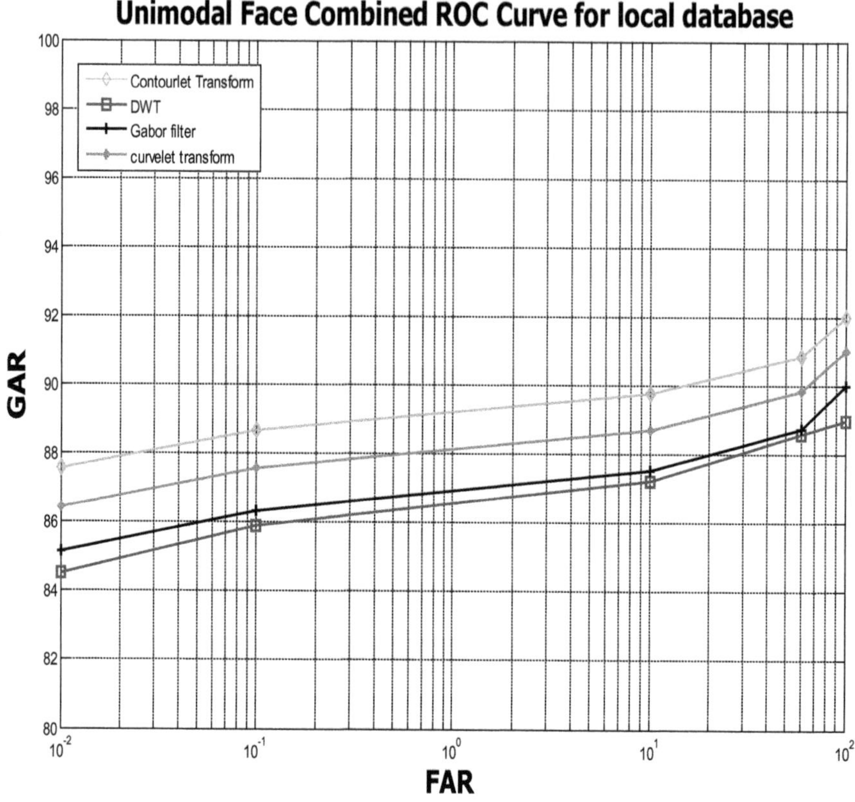

FIGURE 5.19   Unimodal face combined ROC for the local database.

TABLE 5.4
Comparison of different face identification techniques for the CASIA database

Face Methods	FAR	FRR	EER	GAR
DWT	0.8667	0.10	0.82	90
Gabor	0.9895	0.108	0.88	89.1304
Curvelet	0.8667	0.095	0.74	90.5
Contourlet	0.861	0.0798	0.70	92.02

setup that includes a digital camera and web cam. Some biometric features, however, need the data to be collected using a specialised, expensive scanner. Pegs are necessary when scanning the photos in order to obtain precise dimensions. This thesis introduces a novel method for hand recognition that does not require pegs for hand picture acquisition.

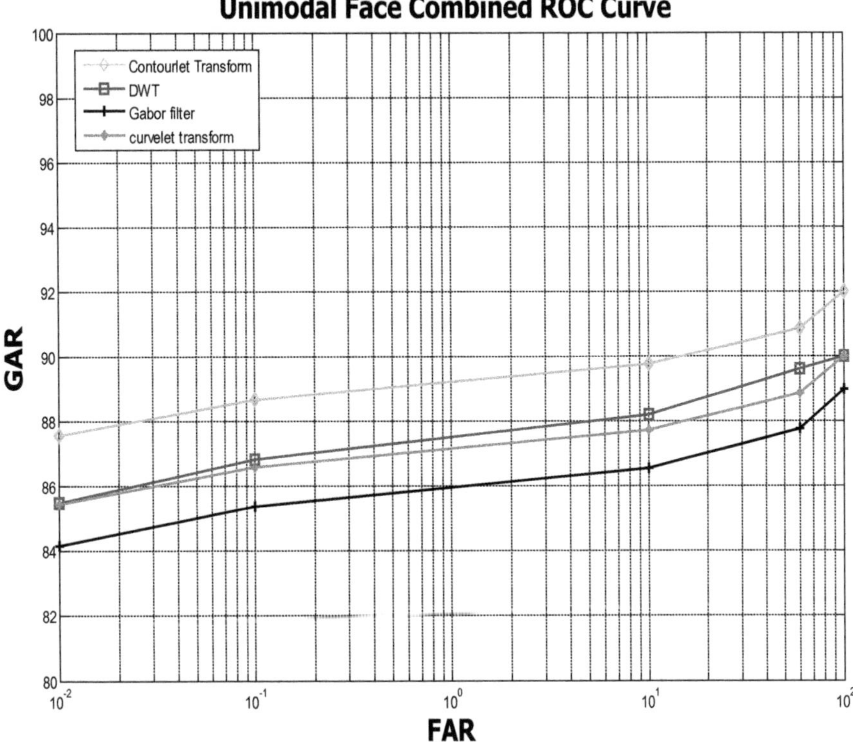

FIGURE 5.20 Unimodal face combined ROC for the CASIA database.

(i) Pre-processing

Image pre-processing is the process of preparing an image for use and analysis at a later time. The pre-processing module's job is to get the picture prepared for feature extraction. The following are the steps in pre-processing:

- Converting RGB to greyscale

To convert an RGB image to a greyscale image, use

$$\text{Intensity} = 0.1140 \times \text{blue} + 0.5870 \times \text{green} + 0.2989 \times \text{red} \qquad (5.7)$$

- Binarisation

The greyscale image will be transformed into a picture with two levels in this step: 0 for white and 1 for black.

Input image (i, j) < threshold, Output image $(i, j) = 1$ Input image $(i, j) \geq$ threshold, Output image $(i, j) = 0$ $G(i, j)$ indicates the grey value of pixel (i, j) after binarisation, $I(i, j)$ represent the original grey value.

$$if\ G(I, j) = \begin{cases} 1, & I(i, j) > threshold \\ 0, & otherwise \end{cases}$$

(ii) Feature extraction

To obtain boundary of hand shape, the following algorithm is used

Step 1: First, locate the hand image's bottom left and bottom right points. Finding the bottom left point involves searching from the bottom left pixel to the bottom right pixel, and finding the bottom right point involves searching from the bottom right pixel to the bottom left pixel based on the pixel value. This concludes the search at this point. At the moment, the hand image's bottom, right, and left reference points total is 12.

Step 2: The region border starts from 12 point. In the border tracing, algorithm considered eight directions from 0 to 7. Figure 5.21 shows eight directions in the border tracing algorithm

Step 3: Variable "dir" Set to be 7.

Step 4: Conduct an anticlockwise search of the current pixel 3 × 3 neighbourhood, beginning with the pixel positioned in the direction. Figure 5.22 shows Pre-processing of hand image.

$$\{(dir+7)\ mod\ 8\}, if\ dir\ is\ even$$
$$\{(dir+6)\ mod\ 8\}, If\ dir\ is\ odd$$

The dir. value is changed by the new boundary element, which is the first pixel identified to have the same value as the current pixel.

Step 5: Stop if the current boundary element equals the second border element and the preceding border element, $n-1$, is equal to 0. If not, continue step 2.

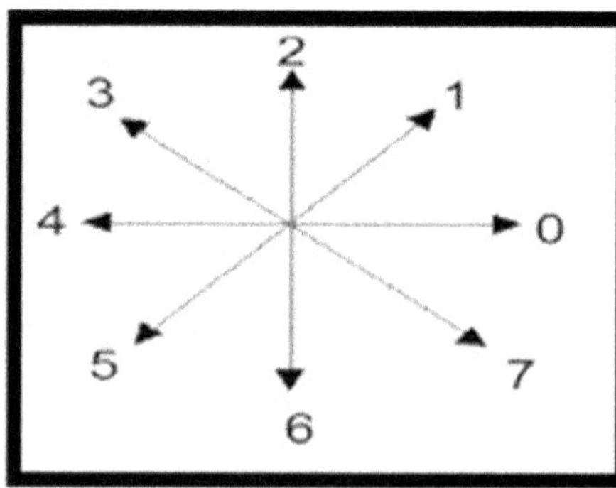

FIGURE 5.21   The eight directions in the border tracing algorithm.

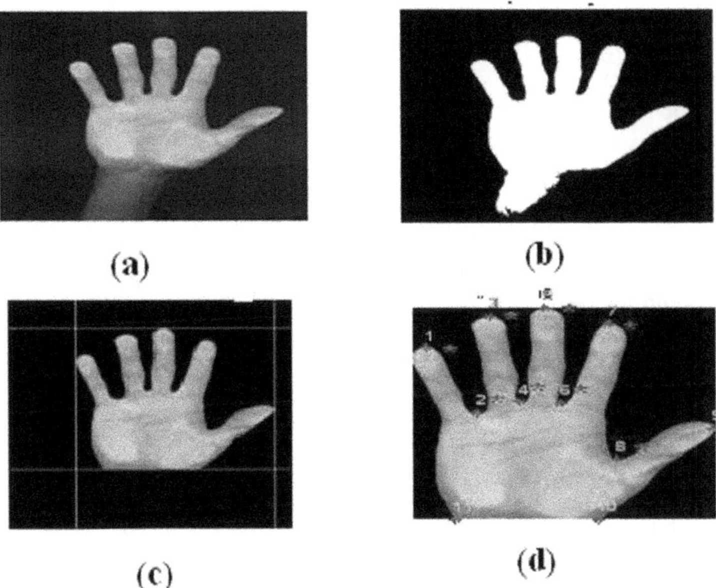

FIGURE 5.22 Results of pre-processing of hand image: (a) original image, (b) binarised image, (c) boundary extracted image, (d) marking of minimum and maximum points on the hand image.

5.4.1 Hand Geometry Recognition Using 12 Geometry Features

In this system, the distance between maxima and minima points is considered to obtain the feature vector. A total of 6 minima points and 6 maxima points are used, so there are a total of 12 distance parameters as shown in Figure 5.23. The feature vector size is 12.

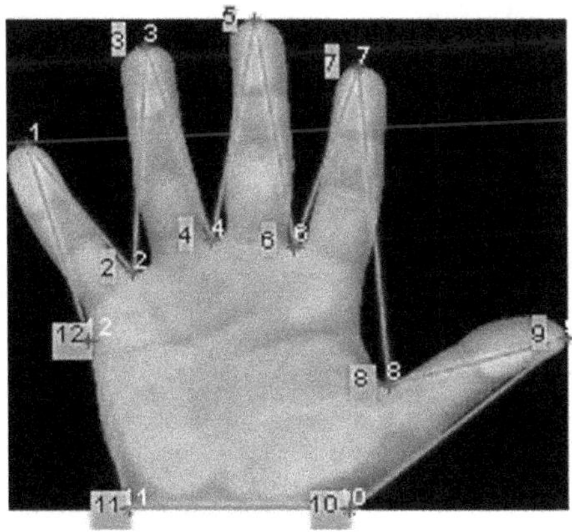

FIGURE 5.23 Hand geometry 12 features.

7 = Index finger Tip point
5 = Middle finger Tip point
3 = Ring finger Tip point
1 = Little finger Tip point
8 = Of index finger Valley point
6 = Between index finger and middle finger Valley point
4 = Middle finger Valley point
2 = Ring finger Valley point
10 = Little finger Valley point
9 = Thumb Valley point
11 = Palm point
12 = Palm bottom left point
13= Palm bottom right point

5.4.2 HAND GEOMETRY RECOGNITION USING 21 GEOMETRY FEATURES

Each finger is measured using one length and three width measurements in this approach. In addition, the measurement of the palm's width and the four other lengths between the thumb valley points and the index, ring, middle, and little valley points. Four fingers yield a total of 16 attributes, one of which is a finger length.

Each finger has three finger widths. Twenty-one features total—four palm widths and the distance between the thumb valley point and the other valley points—are displayed in Figure 5.24.

A total of 21 features are extracted as follows:

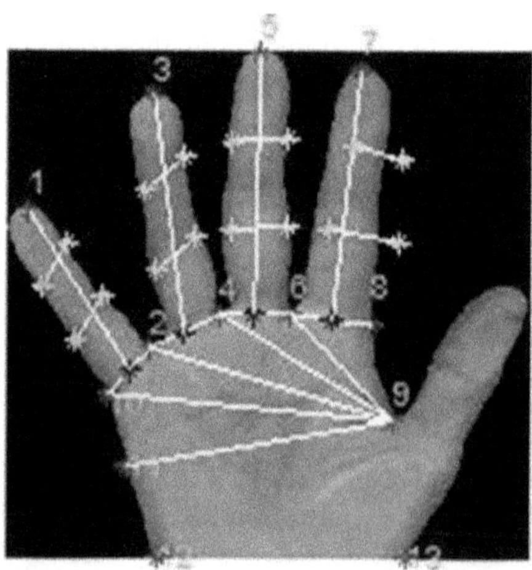

FIGURE 5.24 Hand geometry 21 features.

- The distance between the fingertip and the middle point of the finger baseline is used to calculate finger length. Four features are derived from finger length according to the suggested system, which takes one measurement of each finger's length.
- As seen in Figure 5.24, finger width is measured at three separate sites on the finger: the midpoint of the finger's length, one-third of the finger's length, and the separation between the finger's two minima points.
- Figure 5.24 depicts the palm width as the distance between 9 and 11.
- The separation between the thumb valley point and the tiny valley point is D1 (9–10).
- The distance D2 (9–2) is the separation between the thumb and ring finger valley points.
- The distance between the middle finger's valley point and the thumb valley point is D3 (9–4).
- The distance between the thumb valley point and the index valley point is D4 (9–6).

Table 5.5 shows the result of unimodal hand geometry identification system for the local database. The EER value of 21 features is less as compared to 12 features of hand geometry. Performance of 21 features is better as compared to 12 features of hand geometry.

Figure 5.25 shows the graph of ROC curve (FAR vs GAR), which gives the result of unimodal hand geometry identification system for the local database. The GAR value of 21 features is more as compared to 12 features of hand geometry. The performance of 21 features is better as compared to 12 features of hand geometry.

Table 5.6 shows the result of unimodal hand geometry identification system for the CASIA database. The EER value of 21 features is less as compared to 12 features of hand geometry. The performance of 21 features is better as compared to 12 features of hand geometry.

Figure 5.26 shows the graph of ROC curve (FAR vs GAR), which gives the result of the unimodal hand geometry identification system for the CASIA database. The GAR value of 21 features is more as compared to 12 features of hand geometry. The performance of 21 features is better as compared to 12 features of hand geometry.

TABLE 5.5

Comparison of hand geometry identification system for the local database

Biometric modality	Features	FAR	FRR	GAR	EER
Hand geometry	12	0.94	0.2	80	0.752
Hand geometry	21	0.871	0.12	88	0.531

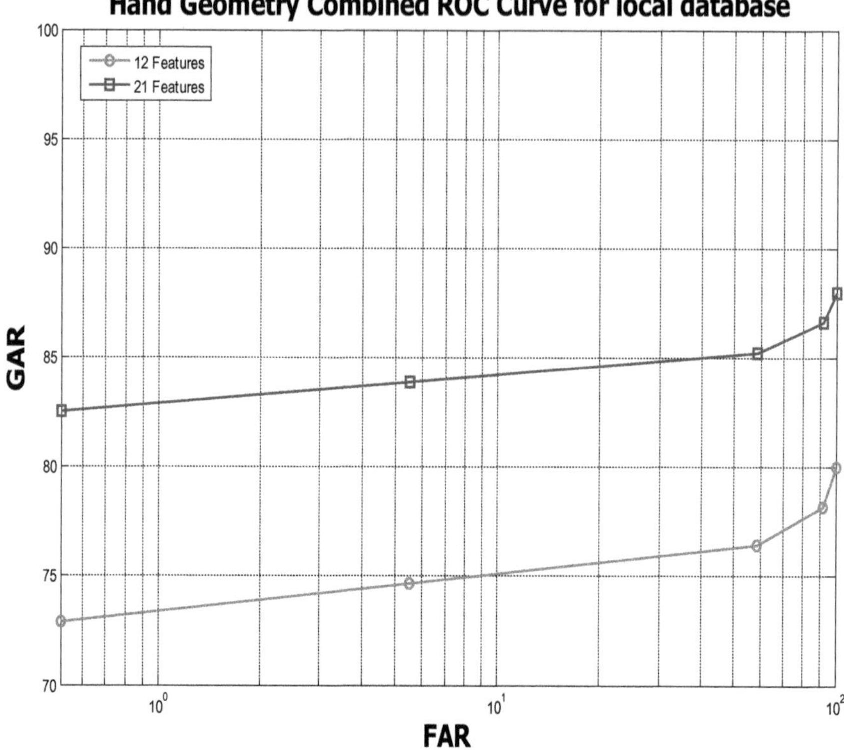

FIGURE 5.25   Unimodal hand geometry combined ROC curve for the local database.

TABLE 5.6

Comparison of hand geometry identification systems for the CASIA database

Biometric modality	Features	FAR	FRR	GAR	EER
Hand geometry	12	0.8754	0.185	81.5	0.5212
Hand geometry	21	0.8517	0.10	90	0.38885

5.5   PALMPRINT AS A BIOMETRIC MODALITY

A fingerprint's texture information provides consistent, distinctive, and repeatable characteristics that are utilised for individual identification. Wavelet transform-based approach is used to extract palm texture feature.

General steps in palmprint recognition:

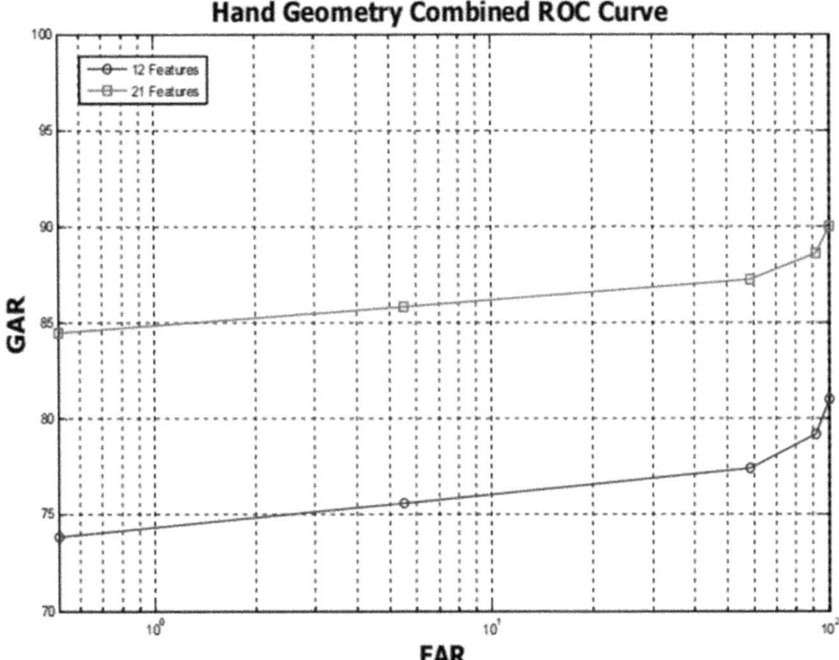

FIGURE 5.26 Unimodal hand geometry combined ROC curve for the CASIA database.

- *Image capture*

  By using digital camera capture the palm images of the person.
- *Image cropping*

  To get the actual palmprint, steps used include thresholding, RGB to grey, edge detection and finding maxima and minima points.
- *Feature extraction*

  After detection of ROI, different features such as principal line, appearance, and texture can be extracted and stored in the database.
- *Recognition*

  The stored features in the database are then compared with the extracted features of input image and the result will display "Yes" or "No".

All the steps up to finding maximum and minimum points are the same as the hand geometry recognition system. For extraction ROI, join two and six minimum points. Draw perpendicular line from 2 and 6 of length 128 pixels. Join perpendicular lines to get square of 128 × 128 pixels. Figure 5.27 shows extraction of palm print image and Figure 5.28 shows Region of Interest of Palm print image after extraction from full hand image.

Table 5.7 shows a comparison between texture-based palmprint identification systems for the CASIA database. Results show that the EER value of contourlet transform is less as compared to other techniques. It shows that the performance of contourlet transform is better than DWT, Gabor filter, and curvelet transform.

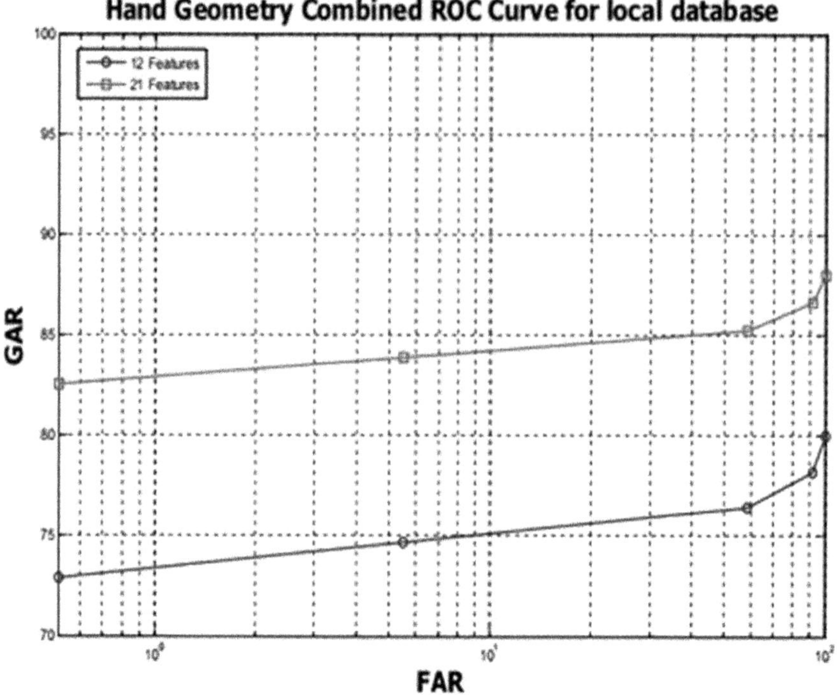

FIGURE 5.27   Extraction of palmprint image.

FIGURE 5.28   ROI of palmprint image.

TABLE 5.7

Comparison of different palmprint identification systems for the CASIA database

Palmprint methods	FAR	FRR	GAR	EER
DWT	0.9334	0.15	85	1.01
Gabor	0.8891	0.1028	89.72	1.1
Curvelet	0.8850	0.1	90	1.09
Contourlet	0.870	0.0627	93.73	0.81

Figure 5.29 shows the graph of ROC curve (FAR vs GAR). It describes the comparison between texture-based palmprint identification systems for the CASIA database. Results show that the contourlet transform GAR value is more as compared to other techniques, which means the performance of contourlet transform is better than DWT, Gabor filter, and curvelet transform.

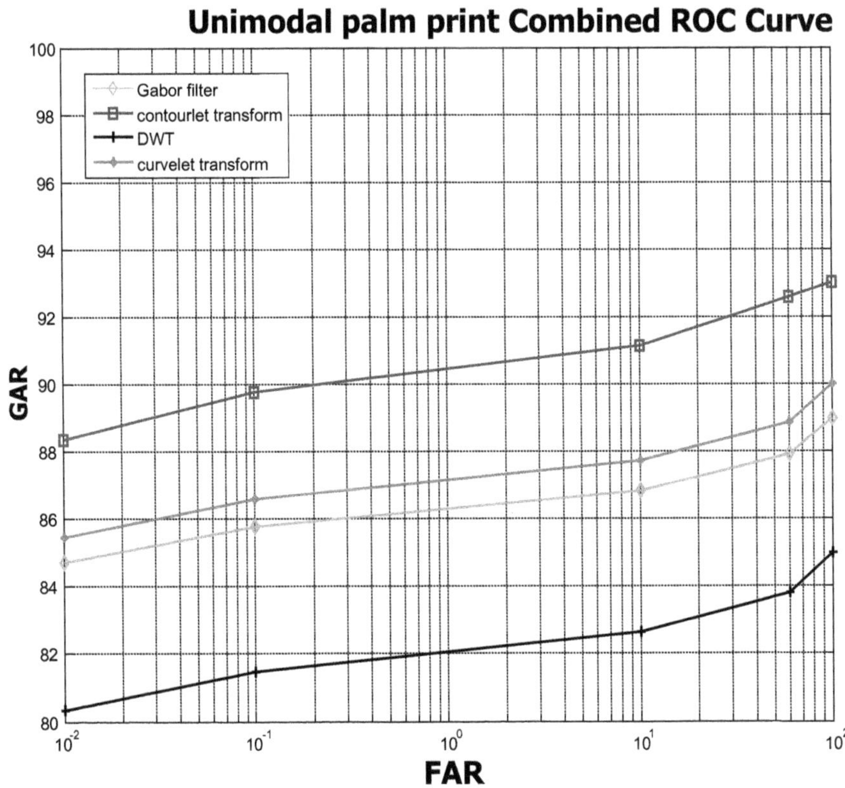

FIGURE 5.29 Graph of the ROC curve (FAR vs GAR).

TABLE 5.8

Comparison of different palmprint identification systems for the local database

Palmprint methods	FAR	FRR	GAR	EER
DWT	1.24	0.13	83	1.24
Gabor	0.942	0.1332	89	1.65
Curvelet	0.901	0.11	89	1.28
Contourlet	0.820	0.0844	91.56	0.93

Table 5.8 shows the comparison between texture-based palmprint identification systems for the local database. Results show that the contourlet transform EER value is less as compared to other techniques which means the performance of contourlet transform is better than DWT, Gabor filter, and curvelet transform.

Figure 5.30 shows the graph of ROC curve (FAR vs GAR). It describes the comparison between texture-based palmprint identification systems for the local database.

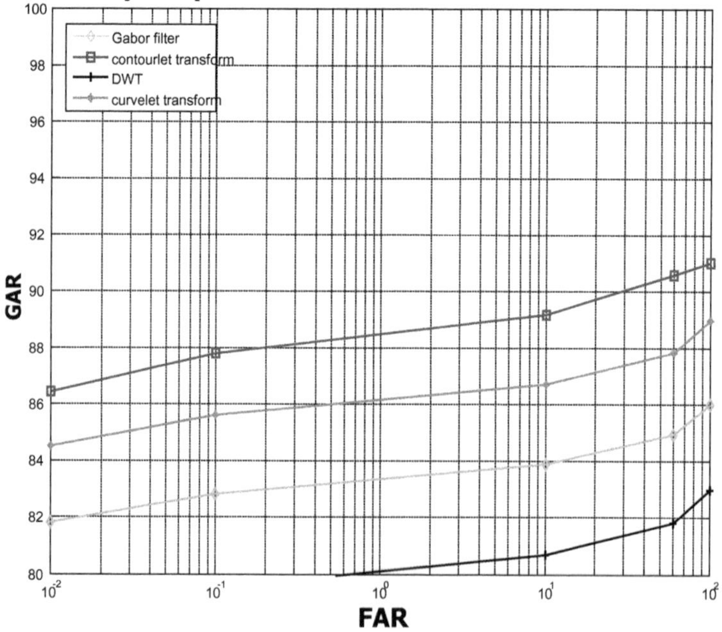

FIGURE 5.30   Palmprint unimodal combined ROC for the local database.

5.5.1 Contourlet Transform

Only horizontal, vertical, and diagonal directional features are provided by DWT. Unequal curve discontinuities are not efficiently handled by the wavelet. Retrieval speed is limited by the Gabor filter's huge feature vector size and considerable computational overhead associated with feature extraction. Although it converges into a continuous domain, the curvelet transform effectively produces curve discontinuities. Contourlet transform starts with a discrete domain and then converge to a continuous domain. Because of its directionality and anisotropy, contourlet transform is superior than wavelet and curvelet transform in describing picture features including lines, edges, contours, and curves].

To represent image properly, its desirable characteristics are as follows:

- Multiresolution
  The image should be successively approximated from coarse to fine resolution.
- Localisation
  The basis element should be localised in both spatial and frequency domains.
- Critical sampling
  For compression, representation should form the basis function with small redundancy.
- Directionality
  The basis element should represent a variety of directions.
- Anisotropy
  To capture smooth contour, basis element should offer extended shapes with diverse aspect ratios.

Although the contourlet transform provides all of these qualities, the first three are supplied by the separable 2D wavelet. A multiresolution directional tight frame known as a contourlet is created to effectively mimic images consisting of smooth regions divided by smooth boundaries. Contourlet transform is built by combining two decomposition phases, and multiscale decomposition followed by directional decomposition. A set of Laplacian sub-bands and a coarse level image can be obtained through multiscale decomposition using the Laplacian pyramid. Critical downsampling at the directional stage allows for additional partitioning into a variable number of frequency sub-bands. The contourlet expansion consists of adjustable aspect ratio basis functions oriented in many directions at different scales. The contourlet transform successfully captures smooth contours, which are the main features of photographs because of its broad collection of basis functions.

At each level, the Laplacian pyramid decomposition creates a down-sampled lowpass version of the original and the difference between the original and the prediction, resulting in a band-pass image, as shown in Figure 5.31. "H" and "G" in this image stand for analysis and synthesis filters, respectively, and "M" represents the sampling matrix. On the coarse version, the procedure can be repeated.

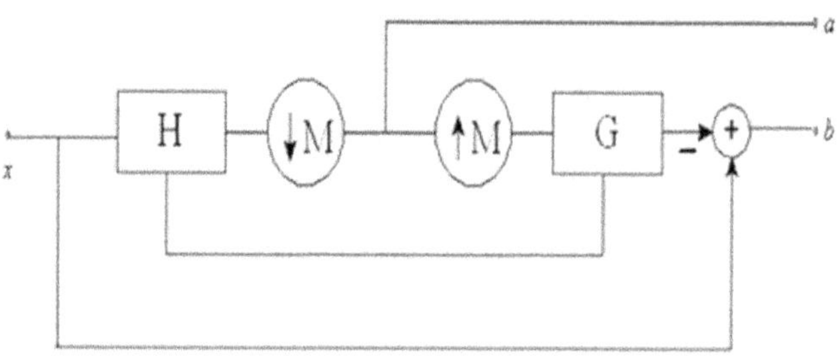

FIGURE 5.31   Laplacian pyramid scheme.

where a = coarse approximation
b = difference between the original signal and the prediction.

The Laplacian pyramid is a collection of band-pass filters as a result. These actions are repeated multiple times to produce a series of photos. A critically sampled filter bank that can break down images into any power of two possible directions is called a directional filter bank (DFB). The DFB is efficiently accomplished by a 1-level tree-structured decomposition that results to "2l" sub-bands with wedge-shaped frequency partition as illustrated in Figure 5.32.

The DFB performs badly when handling low-frequency components. A sparse representation of images is not possible with DFB because low frequencies would leak into multiple directional sub-bands. In order to remedy this, low frequencies are eliminated using LP before applying DFB. When band-pass photos are supplied into a DFB, directional information is effectively recorded.

As shown in Figure 5.33, the number of directions in the DFB doubles when the LP's size is reduced by four times.

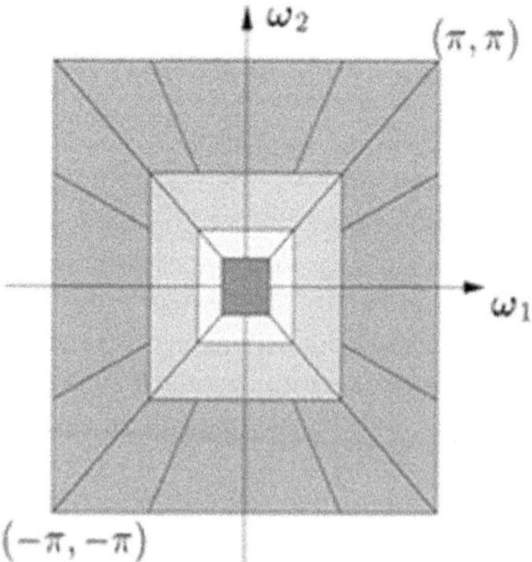

FIGURE 5.32   DFB frequency partitioning.

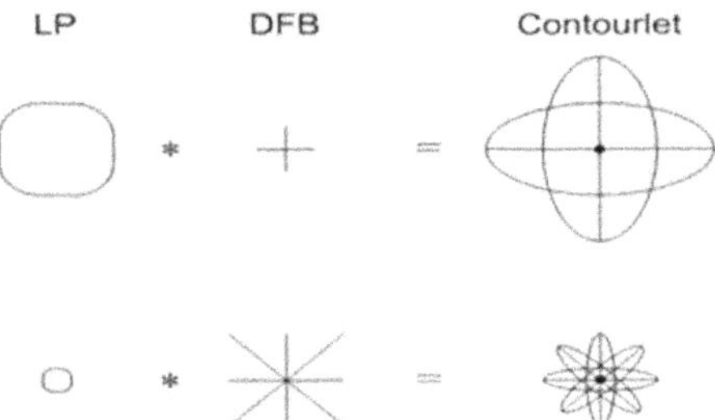

FIGURE 5.33   Contourlet directions.

The experimented results show that for all biometric techniques, contourlet gives better accuracy than DWT, Gabor filter, and curvelet transform. So for fusion, contourlet transform is used to get more accuracy. After applying contourlet transform, mean, energy, variance, PCA, and combination of all features are extracted.

Figures 5.34 and 5.35 display the contourlet transform at level 4, with 0, 2, 3, and 4 directions corresponding to coarse to fine levels. The input image is divided into $2n$ sub-bands at each resolution level "k", where "n" is the filter order. The actual size of the supplied image is 256×256 at the first resolution level, which is the highest resolution. 128×128 is the next resolution level image size. Similar to this, level 3 and level 4 subsampling further reduces the input image, resulting in 64×64 and 32×32 sized images.

After contourlet transform calculated mean, variance, energy, PCA, and a combination of all features form feature vectors.

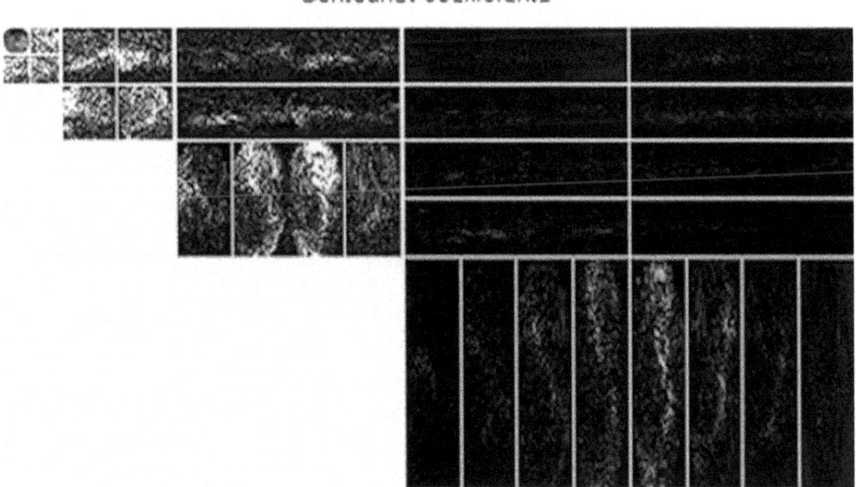

FIGURE 5.34   Contourlet transform decomposition for the fingerprint image.

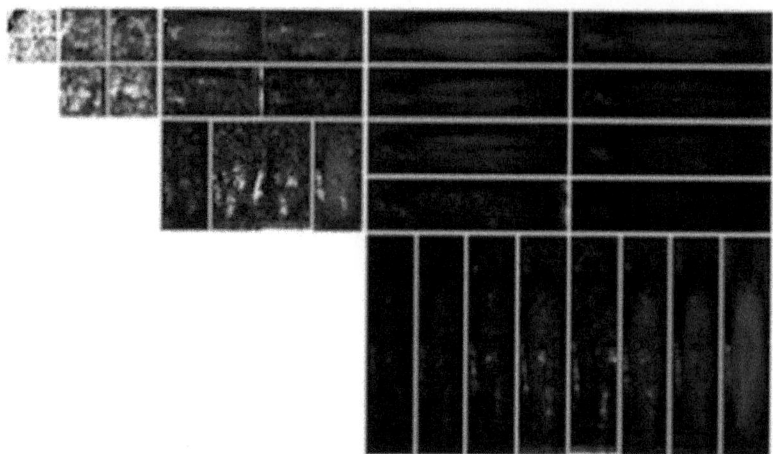

FIGURE 5.35   Contourlet transform decomposition for the face image.

- Mean

    Mean provides a measure of the overall brightness of corresponding image. It has very limited expression and discriminative power.
    Mean=

$$\overline{X} = \frac{\sum_{i=1}^{n} X_i}{n} \tag{5.8}$$

- Variance features

    Variance is the mean squared deviation of each number from its mean. It tells us how far the set of numbers are spread out from mean.

$$\text{Variance} = \sigma^2 = \frac{1}{n} \sum_{i,j=1}^{n} (xij - \overline{x})^2 \tag{5.9}$$

Energy features

    Energy is calculated by an angular second moment. It returns the sum of squared element in a grey-level occurrence matrix. Energy will be "1" for the constant image.

$$\text{Energy} = \sum_{i=1}^{k} \sum_{j=1}^{k} (xij) \tag{5.10}$$

- PCA

PCA is used to calculate a linear transformation that maps data from a higher dimensional space to a low-dimensional space.

PCA algorithm

Suppose $x_1, x_2, ..., x_M$ are Nx_1 vectors.

$$\text{Step} 1 : \overline{X} = \frac{1}{M} \sum\nolimits_{i=1}^{M} X_i \tag{5.11}$$

$$\text{Step} 2 : \text{subtract the mean} : \varnothing_i = xi - x \tag{5.12}$$

$\text{Step} 3 : \text{Form the matrix } A = [\varnothing_1, \varnothing_2, ... \varnothing_M,] (N \times M), \text{then compute}$

$$C = \frac{1}{M} \sum\nolimits_{i=1}^{M} \varnothing_n \varnothing_n^T AA^T \tag{5.13}$$

(Sample covariance matrix, $N \times n$, characterises the scatter of the data)

Step 4: Compute the eigenvalues of C: $\mu_1, \mu_2, ..., \mu N$.

Since C is symmetric, $\mu_1, \mu_2, ..., \mu N$ form a basis (i.e., many vector x or actually $(x\text{-}x\bar{\ }1)$, can be written as a linear combination of the eigenvectors)

$$X - \overline{X} = b1u1 + b2u2 + ...bNuN \tag{5.14}$$

Principal component analysis (PCA) is a statistical technique that transforms a series of observations of potentially correlated variables into a set of values of linearly uncorrelated variables, or principal components, using orthogonal transformation.

Here all features mean, variance, energy, and PCA features combined to generate a new feature vector.

5.6 EUCLIDEAN DISTANCE AS A CLASSIFIER

The Euclidean distance algorithm is used for matching. We accept the value if it above the threshold; if not, we reject it. If $p = (p1, p2)$ and $q = (q1, q2)$, then the Euclidean distance is given by

$$d(p, q) = \sqrt{(p1 - q1)^2 + (p2 - q2)^2} \tag{5.15}$$

As quality of extracted texture features are good, a simple classifier Euclidean distance can provide very satisfactory recognition performance. There is no need to employ any complicated classifier.

Table 5.9 represents the results of unimodal face identification techniques using different features after contourlet transform. Experiment confirm the variance as the best feature since EER value is less as compared to mean, energy, PCA, and combination of all features.

Figure 5.36 shows the graph of FAR vs GAR for unimodal face identification techniques using different features after contourlet transform, which shows that the variance features GAR value is more as compared to mean, energy, PCA, and a combination of all features.

TABLE 5.9

Results of unimodal face identification technique using different features after contourlet transform

Biometric modality	Features	FAR	FRR	GAR	EER
Face	Mean	0.81	0.405	59.5	0.4525
Face	Energy	0.805	0.35	65	0.3825
Face	Variance	0.71	0.088	91.2	0.355
Face	PCA	0.805	0.318	68.2	0.4737
Face	All features	0.812	0.09	91	0.5125

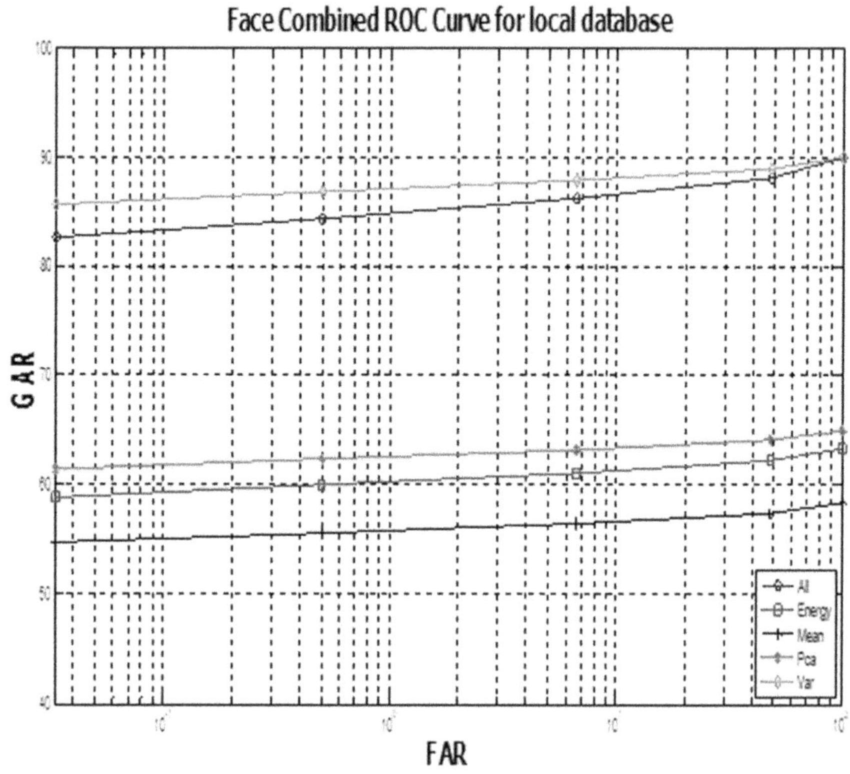

FIGURE 5.36 Unimodal face combined ROC curve using contourlet transform.

Table 5.10 represents the results of unimodal palmprint identification techniques using different features after contourlet transform. Experiment confirm variance as the best feature since the EER value is less as compared to mean, energy, PCA, and combination of all features.

TABLE 5.10

Results of unimodal palmprint identification technique using different features after contourlet transform

Biometric modality	Features	FAR	FRR	GAR	EER
Palmprint	Mean	0.754	0.009	91	0.4212
Palmprint	Energy	0.705	0.05	95	0.48885
Palmprint	Variance	0.754	0.045	95.5	0.2175
Palmprint	PCA	0.613	0.006	94	0.35
Palmprint	All features	0.627	0.005	95	0.445

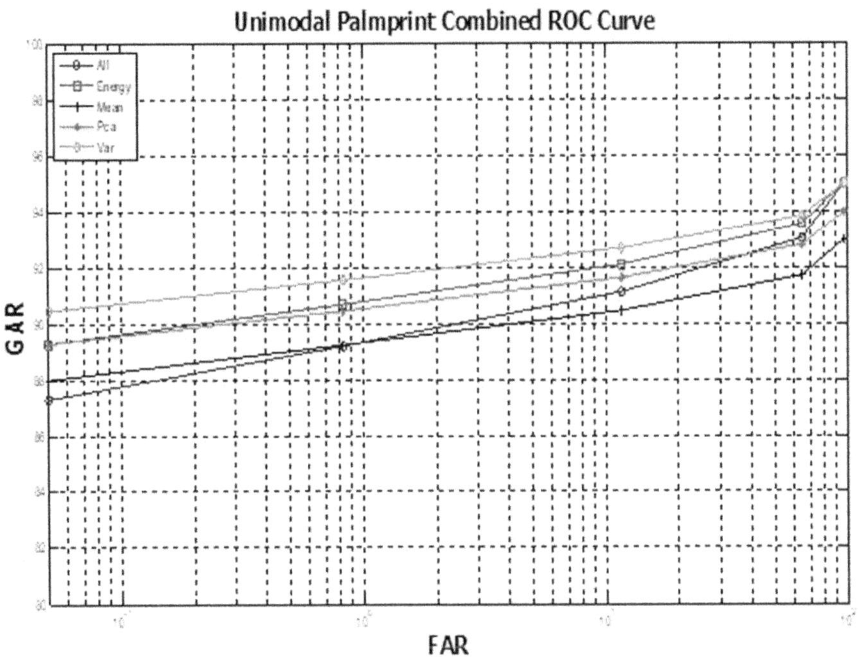

FIGURE 5.37 Unimodal palmprint combined ROC curve using contourlet transform.

Figure 5.37 shows the graph of FAR vs GAR for unimodal palmprint identification techniques using different features after contourlet transform, which shows that the variance features GAR value is more as compared to mean, energy, PCA, and a combination of all features.

Table 5.11 represents the results of unimodal fingerprint identification techniques using different features after contourlet transform. Experiment confirm variance as the best feature since the EER value is less as compared to mean, energy, PCA, and a combination of all features.

TABLE 5.11

Results of unimodal fingerprint print identification technique using different features after contourlet transform

Biometric modality	Features	FAR	FRR	GAR	EER
Fingerprint	Mean	0.86	0.07	93	0.121
Fingerprint	Energy	0.85	0.05	95	0.1642
Fingerprint	Variance	0.786	0.02	98	0.1016
Fingerprint	PCA	0.892	0.065	93.5	0.11
Fingerprint	All features	0.83	0.02	98	0.19

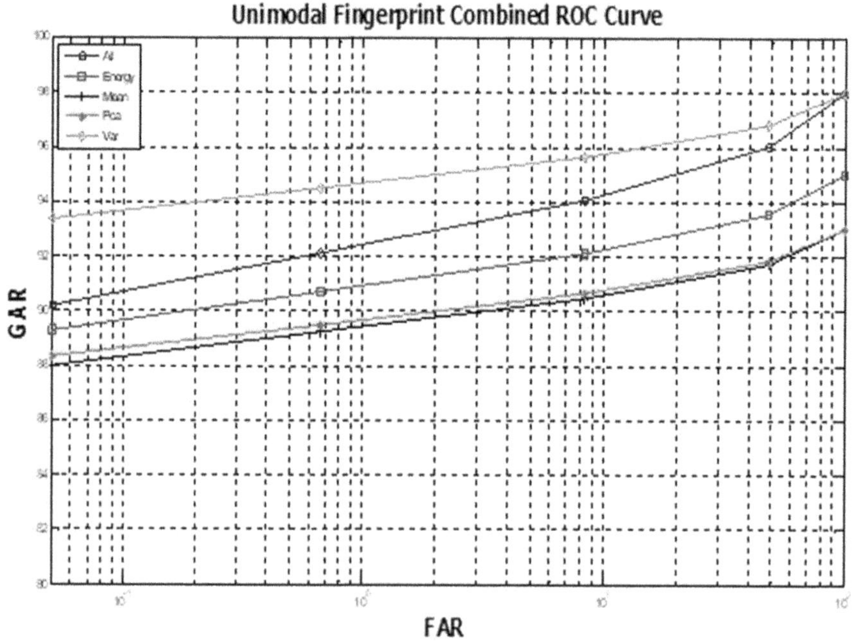

FIGURE 5.38   Unimodal fingerprint combined ROC curve using contourlet transform.

Figure 5.38 shows the graph of FAR vs GAR for unimodal fingerprint identification techniques using different features after contourlet transform, which shows that the variance features GAR value is more as compared to mean, energy, PCA, and a combination of all features.

5.7 SUMMARY

This chapter discusses the unimodal biometric traits which are compared after the feature extraction to test the efficiency of each biometric trait. It also discusses contourlet transform as the best modality-specific feature extraction algorithms for fingerprint, face, and palmprint, respectively, and for dimension reduction of the feature vector, the linear discriminate analysis approach is used.

REFERENCES

1. C-C. Han, H-L. Cheng, C-L. Lin, and K-C. Fan, "Personal authentication using palmprint features", *Pattern Recognition*, vol. 36, 2003, pp. 371–381.
2. Loris Nanni and Alessandra Lumini, "Ensemble of multiple palmprint representation", *Expert Systems with Applications*, 36, 2009, pp.4485– 4490(Elsevier).
3. Jiansheng Chen, Yiu-Sang Moon, Ming-Fai Wong c, and Guangda Su, "Palmprint authentication using a symbolic representation of images", *Image and Vision Computing*, vol. 28, 2010, pp. 343–351, Elsevier journal.

6 Multimodal Biometric Identification Systems Using Sensor-Level Fusion

6.1 MULTIMODAL BIOMETRIC IDENTIFICATION SYSTEM

The system which combines information from several different biometric traits for the identification of a person is called a multimodal biometric identification system.

Biometric identification systems, as a system, have faced many problems such as noise in sensed data, intra-class variations, inter-class similarities, non-universal and spoof attacks, etc. Biometric fusion, especially multimodal biometric fusion, is to overcome the lack of modification and increase the reliability of the system.

Multimodal biometric systems use different levels of fusion to combine two or more transformations: (i) sensor-level fusion, which combines two images from different sensors; (ii) feature extraction-level fusion, where features are extracted using two or more types of modalities (iii) matching score-level fusion, combining scores from various systems; and (iv) integration at the decision level where combinations of multiple accept/reject decisions are made for the system.

Early integration strategies result in better performance than late integration strategies. In this book, two fusion techniques named sensor level and feature level are presented for combining the feature scores of randomly selected biometric traits as shown in Figure 6.1. Biometric characteristics were selected using logistic regression. Therefore, the weight of the four biometric features is given according to their accuracy. The weights assigned to finger, palm, face, and hand geometry are 0.3, 0.25, 0.25, and 0.2, respectively. More weight is given to more accurate biometric features.

6.2 SENSOR-LEVEL FUSION

Sensor-level fusion is carried out immediately after the acquisition of an image from the sensor. The raw biometric sample is the richest source of information since it has been minimally processed. Raw data from various sensors can be processed and fused to create new vectors. These raw data could be multiple acquisitions of the same biometric trait from a single sensor or multiple acquisitions of different traits.

DOI: 10.1201/9781032665993-6

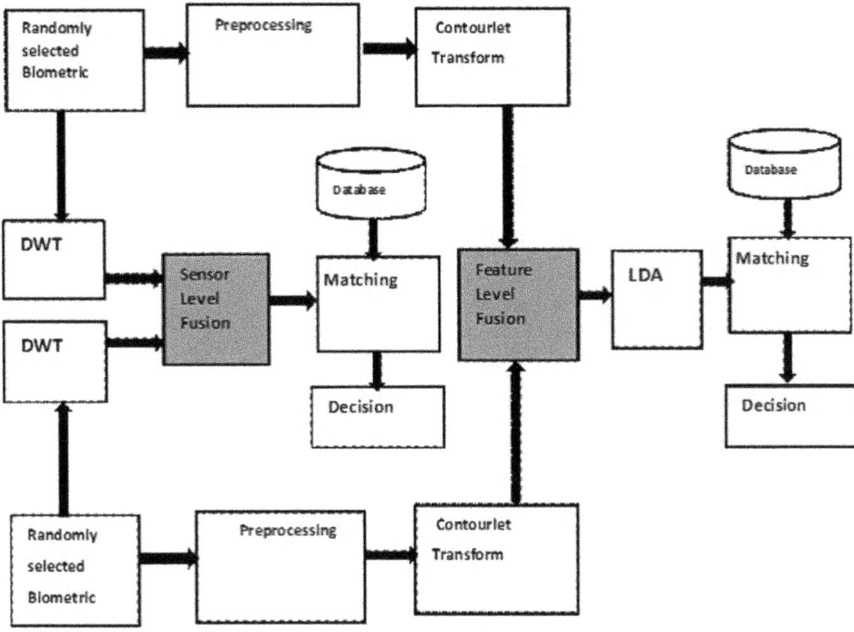

FIGURE 6.1 Multimodal biometric identification system.

In this research, acquired multiple modalities from various sensors are integrated together to get a single biometric modality.

6.3 BASIC STRUCTURE FOR SENSOR-LEVEL FUSION

In sensor-level fusion, detailed information obtained from both source images is fused to get an effective representation of the final image. The standard fusion method gives spectral distortion but performs well spatially. Multiscale transform-based fusion is used to remove such problems. Images obtained from different sources are fused using multiresolution images. Discrete wavelet transform (DWT) is a multi-resolution decomposition which provides an approximation image and three directional images, viz., vertical, horizontal, and diagonal images. This research presents a novel approach for the fusion of different texture information images with the same resolutions at the sensor level using a random selection of biometric traits. The various sensors are used to acquire face, fingerprint, and palmprint images. Discrete wavelet transform and decomposition provide new fused images from two different randomly acquired images which are re-sized before fusion. Most researchers [49, 55, 56] described multiresolution image fusion of the same subject, which are collected from multiple sensors. The presented approach for image fusion of different subjects is collected from different sensors. Before sensor-level fusion, for the decomposition of any two randomly selected biometrics, the DWT is used. The wavelet transform contains fine-level (LL band) and coarse-level images (HL, LH,

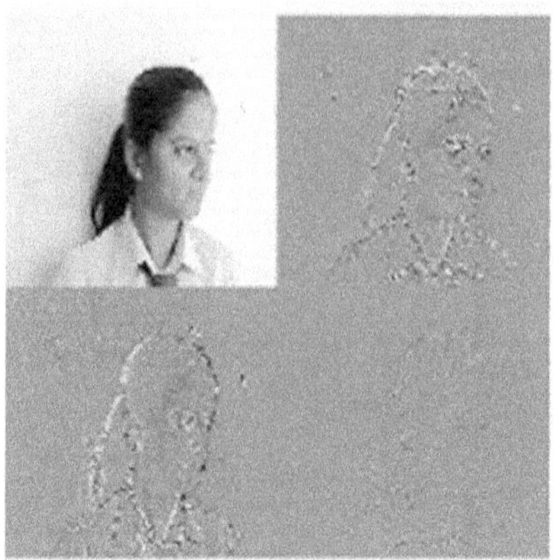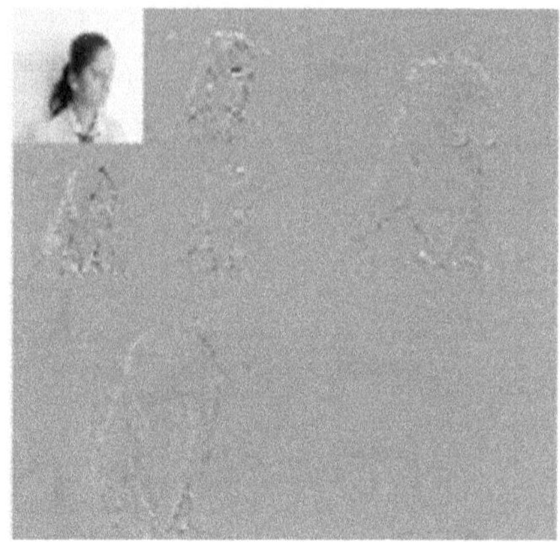

FIGURE 6.2 First- and second-level decomposition of face image.

HH bands) at different scales. The LL band has all positive changing values and the additional band has a zero changing value. A large change in brightness means that important features in the image (such as edges, lines, borders) change, that is, the larger the value, the more it changes. Thus, the larger absolute value of two DWT coefficients at each point is selected for a fusion of two images. To obtain a new fused image, an inverse wavelet transform is performed. The wavelet transform iteratively decomposes the image into multiple frequencies, each with a variable value.

Figures 6.2 and 6.3 show the first-level and second-level decomposition of DWT for face and palmprint image, respectively. Figure 6.4 shows the sensor-level fusion of face and palmprint images.

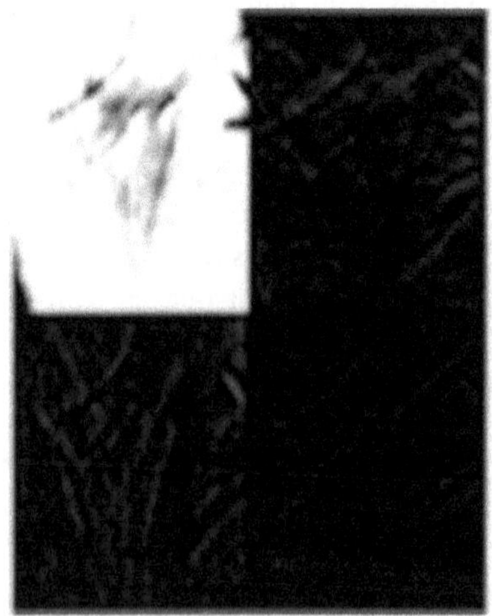

FIGURE 6.3 First- and second-level decomposition of palmprint image.

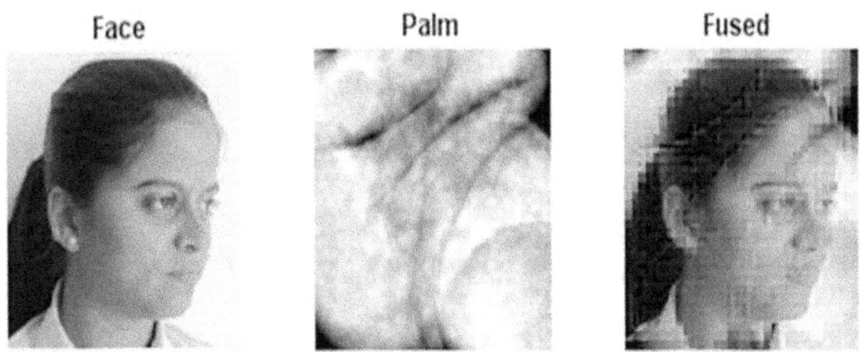

FIGURE 6.4   Sensor-level fusion of face and palmprint image.

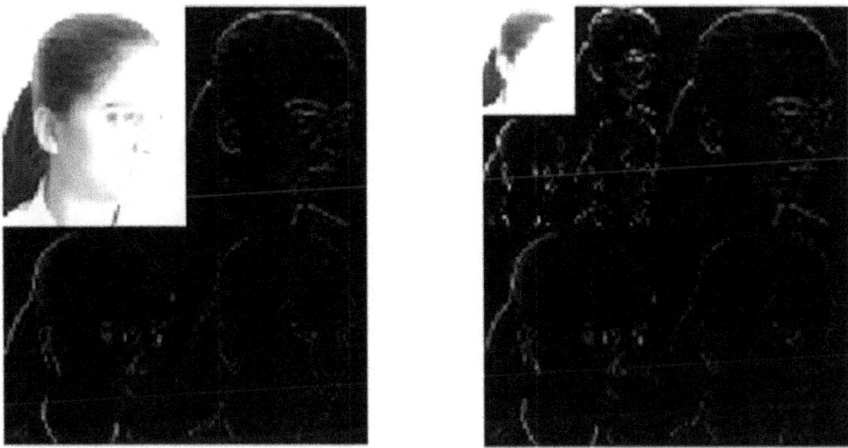

FIGURE 6.5   First- and second-level decomposition of face image.

Figures 6.5 and 6.6 show the first-level and second-level decomposition of DWT for face and fingerprint images, respectively. Figure 6.7 shows the sensor-level fusion of face and fingerprint images.

Figures 6.8 and 6.9 show the first-level and second-level decomposition of DWT for fingerprint and palmprint images, respectively. Figure 6.10 shows the sensor-level fusion of fingerprint and face images.

6.4   SENSOR-LEVEL FUSION OF LOW-FREQUENCY AND HIGH-FREQUENCY FEATURES

DWT is calculated with continuous low-pass and high-pass filtering which provides one fine image and three detail images. After each transform pass, at the next lower resolution level, the low-pass frequency again generates high- and low-frequency sub-bands. This research uses the second-level Daubechies transform for the fusion

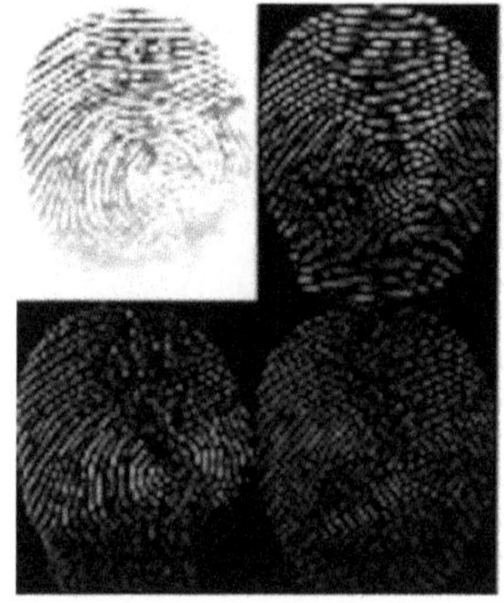 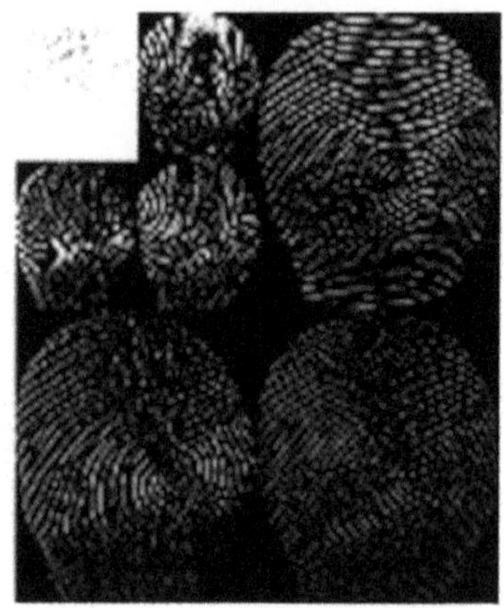

FIGURE 6.6   First- and second-level decomposition of fingerprint image.

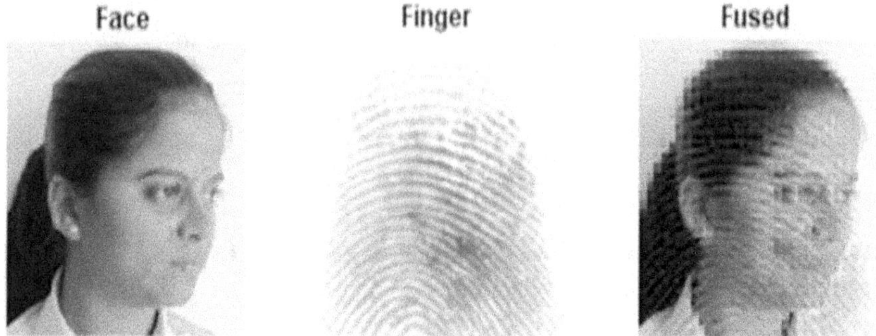

FIGURE 6.7   Sensor-level fusion of face and fingerprint image.

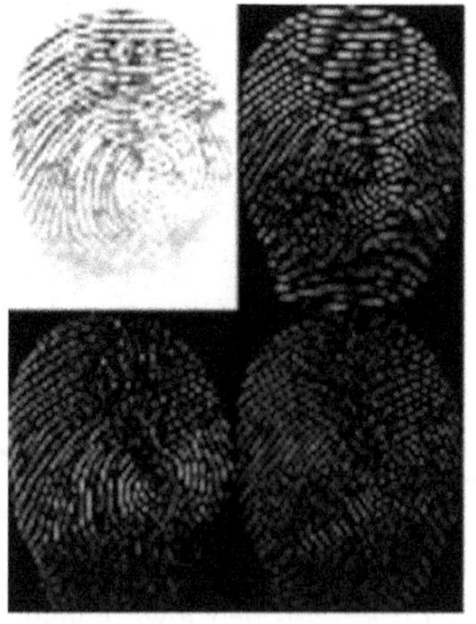 

FIGURE 6.8   First- and second-level decomposition of fingerprint image.

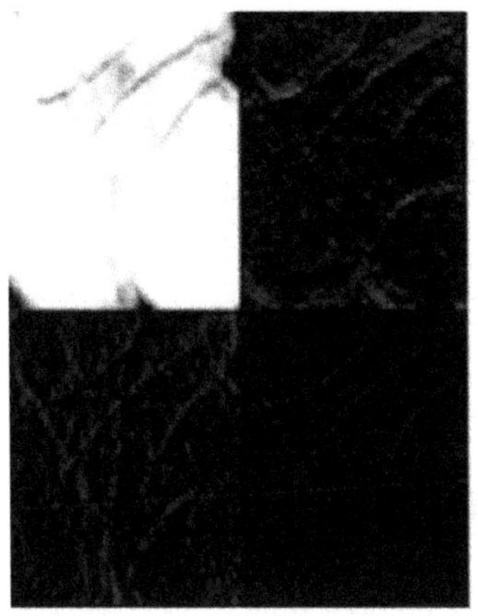 

FIGURE 6.9 First- and second-level decomposition of palmprint image.

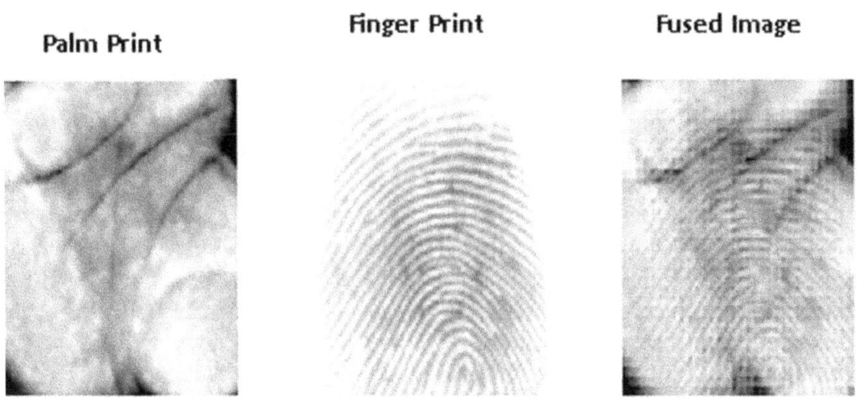

FIGURE 6.10 Sensor-level fusion of palmprint and fingerprint image.

of low-frequency and high-frequency features. For image fusion, the maximum values of low- and high-frequency coefficients are used. By means of the inverse wavelet transform $w-1$, the output fusion image is obtained.

After the second-level decomposition, the image would be

$$I = I_{LL2} + I_{LH2} + I_{HL2} + I_{HH2} \tag{6.1}$$

LH2, *HL2*, and *HH2* represent the vertical, horizontal, and diagonal transform values. This provides high-frequency points where the transform values fluctuate around zero. The *LL2* low-frequency content has a positive transform value. Perform the inverse wavelet transform as shown in Figure 6.11 to obtain the final merged

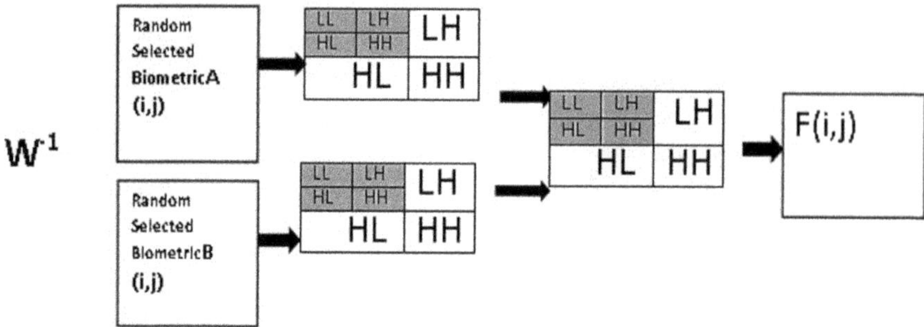

FIGURE 6.11 Sensor-level low- and high-frequency feature fusion.

image. For randomly selected biometric images, the merged images provide good quality of image information.

The input image size is 256 × 256. After second-level decomposition, it becomes 128 × 128. So the feature vector size is 128 × 128. For the fusion, the maximum grayscale values of the first and second modalities are considered.

$$W = \max [A\,(i, j), B\,(i, j)] \tag{6.2}$$

6.5 SENSOR-LEVEL FUSION OF LOW-FREQUENCY FEATURES

At different scales, discrete wavelet transform contains fine level (LL band) frequency bands which have positive transform values and coarse level (LH, HH, HL bands) frequency bands wherein transform values are fluctuating around zeros. More changes in brightness mean changes to key features in the image, such as edges, lines, and borders, resulting in higher transform values. The presented fusion is of only low-frequency features of any two randomly selected biometrics. For image fusion, the maximum values of the coefficients obtained by the LL band of each image are considered.

The fusion rule consists

$$F\,(i, j) = \text{maximum}\,(LL1, LL2) \tag{3.18}$$

The input image size is 256 × 256. After second-level decomposition, it becomes 128 × 128. The approximated image size is 64 × 64. For fusion, the maximum grayscale values of first and second modalities are considered as shown in Figure 6.12.

Table 6.1 shows a combination of low-frequency features and high-frequency features, which gives more accuracy as compared to a combination of only low-frequency features. A low frequency provides positive transform values and a high frequency detail provides fluctuating transform values around zeros.

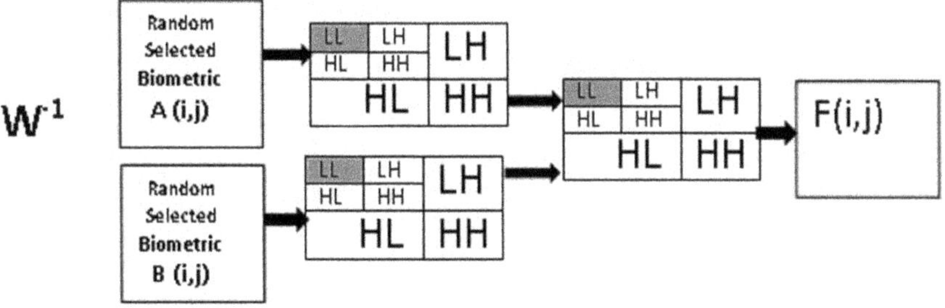

FIGURE 6.12 Sensor-level low-frequency feature fusion.

TABLE 6.1

Comparison of different biometric modalities at sensor-level fusion

Biometric modality	Features	FAR	FRR	GAR	EER
Face_fingerprint	64 × 64	0.628	0.0117	98.8	0.1503
Face_fingerprint	128 × 128	0.3563	0	100	0.107
Face_palmprint	64 × 64	0.013	0.022	97.8	0.19
Face_palmprint	128 × 128	0.312	0	100	0.04
Palm_fingerprint	64 × 64	0.31	0.0163	98.4	0.0286
Palm_fingerprint	128 × 128	0.257	0	100	0.0405

Figure 6.13 shows the graph of ROC curve (FAR vs GAR) for sensor-level fusion of fingerprint and palmprint for combining low-frequency and both low-frequency and high-frequency features. The GAR value is higher for a combination of low- and high-frequency features than a combining only low-frequency features Figure 6.14 shows the graph of ROC curve (FAR vs GAR) for sensor-level fusion of Face and palmprint for combining low-frequency and both low-frequency and high-frequency features. The GAR value is higher for a combination of low- and high-frequency features than a combining only low-frequency features.

combining only low frequency features.

Figure 6.15 shows the graph of the ROC curve (FAR vs GAR) for sensor-level fusion of fingerprint and face for combining low-frequency features and both low-frequency and high-frequency features. The GAR value is higher for a combination of low and high frequency than a combination of only low-frequency features.

Incompatibility of the information content or heterogeneity is responsible for fusion at the sensor level. For example, in face and hand geometry multimodal system, the fusion of raw images may not be possible due to the heterogeneity of features obtained from face and hand geometry. So fusion of hand geometry features

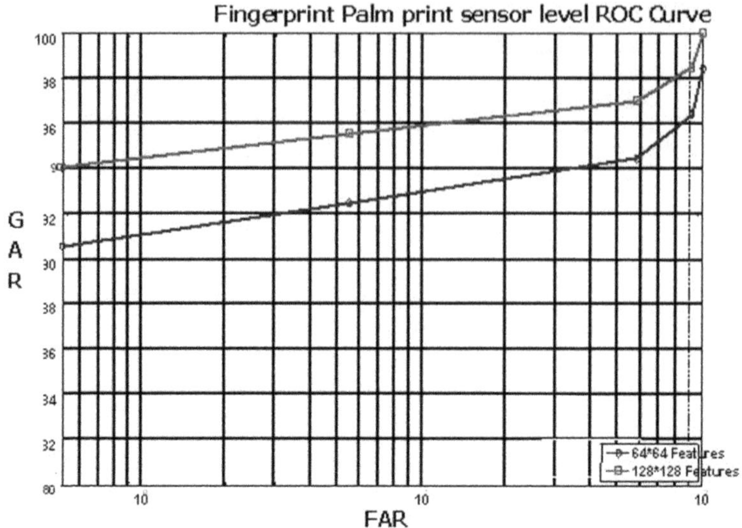

FIGURE 6.13   Fingerprint and palmprint sensor-level fusion ROC curve.

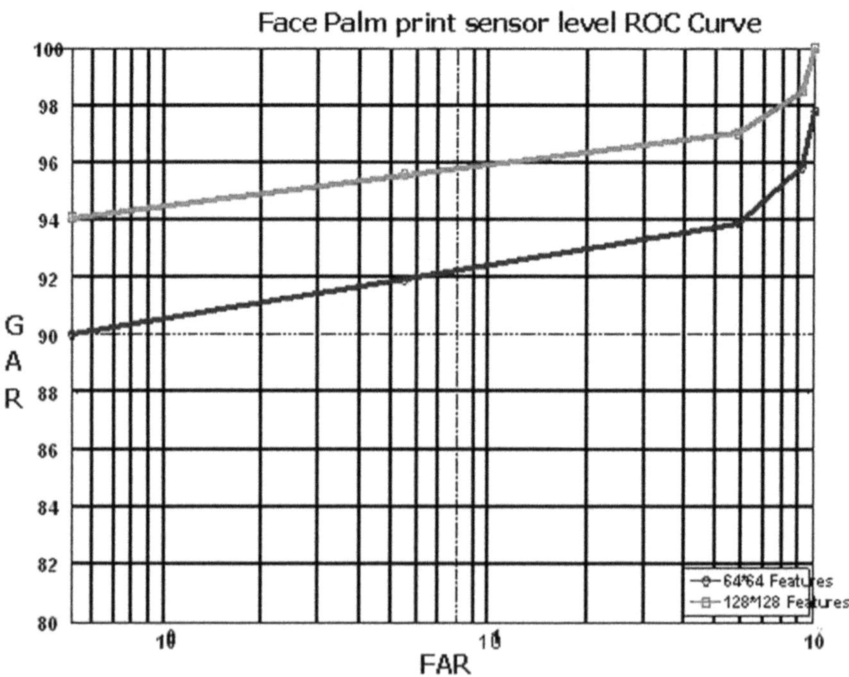

FIGURE 6.14   Face and palm print sensor level fusion ROC curve.

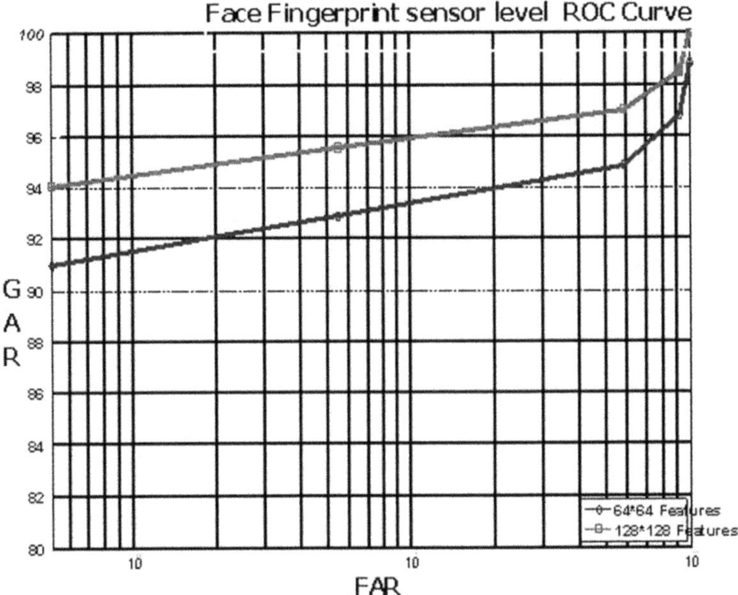

FIGURE 6.15   Face and fingerprint sensor level fusion ROC curve.

with face, palm, and fingerprint is not possible due to the heterogeneity of features. Due to these challenges of sensor-level fusion, this research presented feature-level fusion.

6.6 SUMMARY

Early integration strategies result in better performance than late integration strategies. For sensor-level fusion, features are obtained by combining only low-frequency features and by combining both low- and high-frequency features after applying DWT. For sensor-level fusion, a combination of low frequency and high frequency obtained the best result as compared to only combined low-frequency features.

7 Multimodal Biometric Identification Systems Using Feature-Level Fusion

7.1 MULTIMODAL BIOMETRIC SYSTEM

Multimodal biometric systems use different levels of fusion to combine two or more biometric traits: (i) sensor-level fusion, which combines two images from different sensors; (ii) feature extraction-level fusion, where features are extracted using two or more types of fusion and concatenated to get final feature vector; (iii) matching scores level fusion, combining scores from various matchers; (iv) integration at the decision level where combinations of multiple accept/reject decisions are made for the system. Early integration strategies are more effective than late integration strategies. This book shows how to integrate. Fusion is done through various integration techniques.

In the feature-level fusion process, features are extracted from different biometric traits. Then, the extracted features can be combined to create the final feature vector. Compared with score level and decision-level fusion, feature-level fusion provides better information because the feature set contains more information about the input data. In feature-level fusion, new vectors are obtained by combining vectors obtained from different biometric modality. The dimension of the new feature vector is high which has more feature information on decision requirement. To reduce the dimension of the feature vector, a linear discriminate analysis is used.

In feature-level fusion as shown in Figure 7.1, the feature set obtained from two different sensors (for example, face and fingerprint images are acquired using different sensors) are first pre-processed and then extracted features from each sensor independently to form a joint feature vector. The new single feature vector is formed by concatenating features.

This book presented a fusion of fingerprint, face palmprint, and hand geometry images at the feature level. Features are extracted by using two different methods—block variance features and contourlet transform features—as shown in Figure 7.2.

Figure 7.3 shows the concatenation of two feature vectors obtained from two different biometric modalities.

DOI: 10.1201/9781032665993-7

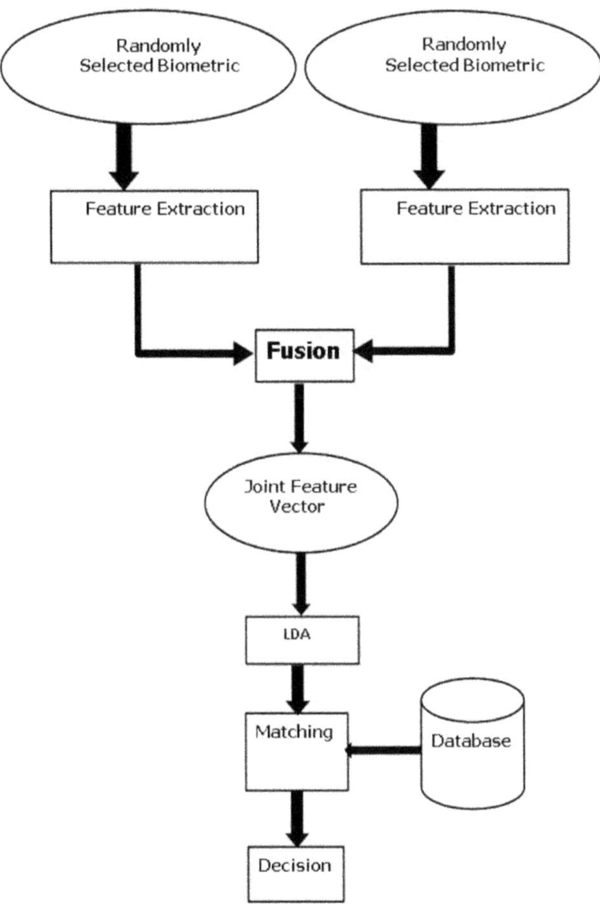

FIGURE 7.1 Block diagram of feature-level fusion.

7.2 FEATURE-LEVEL FUSION USING BLOCK VARIANCE FEATURES

7.2.1 FEATURE-LEVEL FUSION OF 128 FEATURE VECTOR

The feature vector of 128 features is formed by concatenated features of 64 features of the first modality and 64 features of the second modality are used. Input image size is 256 × 256. The image is divided into 32 × 32 windows, which forms 8 blocks in rows and 8 blocks in columns so a total of 64 features are extracted from first modalities and 64 features from second modalities. Concatenated features of both modalities give a total of 128 size of feature vector.

Figure 7.4 shows how the image is divided into 64 blocks which are formed by dividing the image into 8 rows and 8 columns.

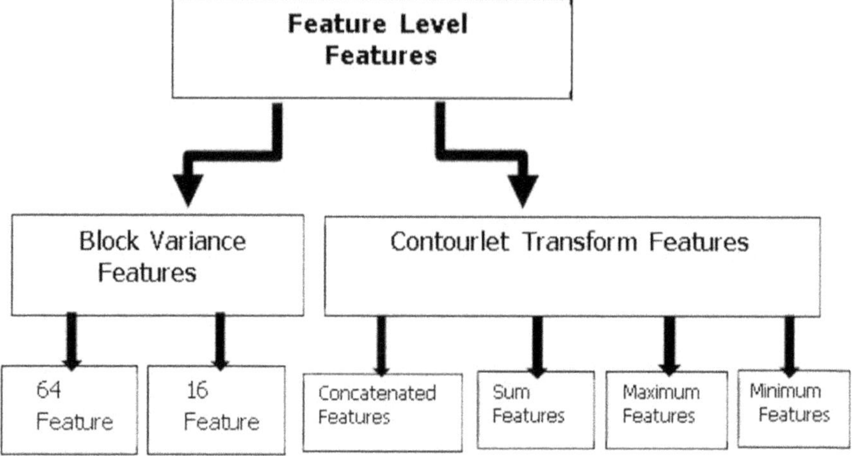

FIGURE 7.2   Different features used for feature-level fusion.

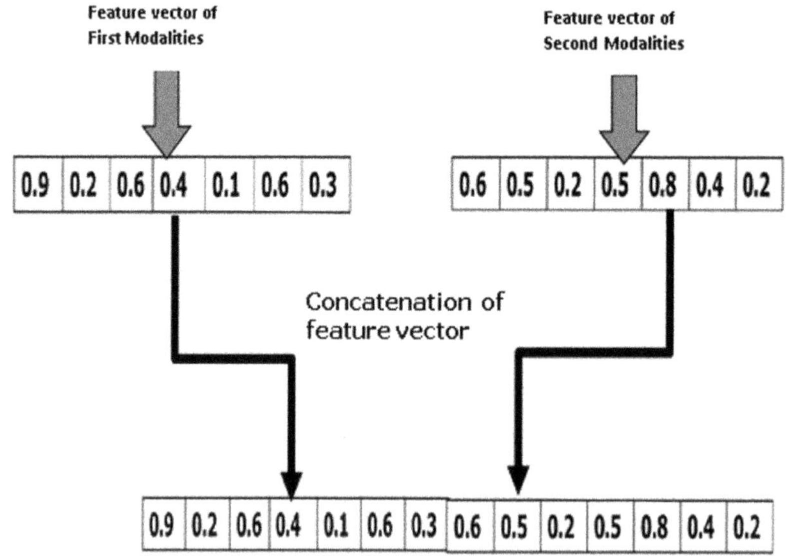

FIGURE 7.3   Concatenation of two feature vectors.

Figures 7.5 and 7.6 show 64 feature vector of face and fingerprint, respectively. Figure 7.7 shows the concatenated combined feature vector of face and fingerprint which shows the first 64 features of face biometrics and the next 64 features of fingerprint biometrics so the total size of the feature vector after concatenation is 128. The x-axis shows the feature vector size and the y-axis shows the magnitude of features.

FIGURE 7.4 Dividing input image into 64 blocks.

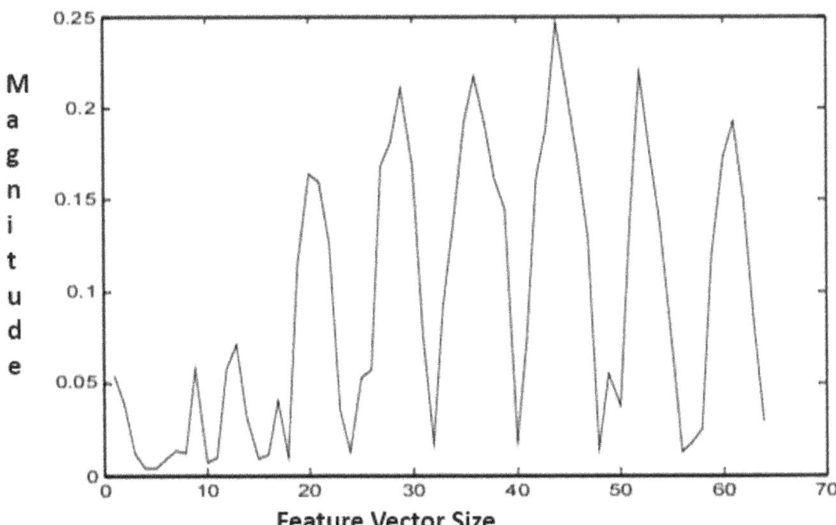

FIGURE 7.5 Face feature vector (64 features).

7.2.2 FEATURE-LEVEL FUSION OF 32 FEATURE VECTOR

To form a feature vector of 32 features, concatenated features of 16 features of the first modality and 16 features of the second modality are used. The input image size is 256 × 256. The image is divided into 64 × 64 windows, which forms 4 blocks in rows and 4 blocks in columns, so a total of 16 features are extracted from first

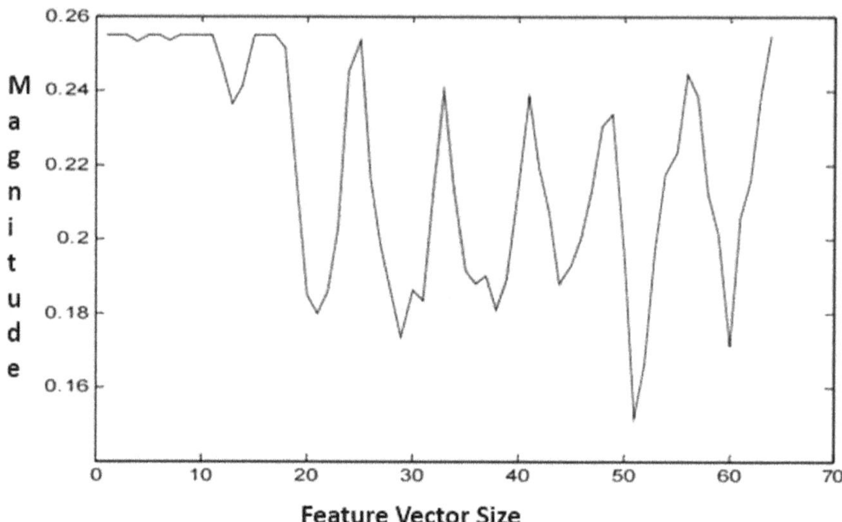

FIGURE 7.6   Fingerprint feature vector (64 features).

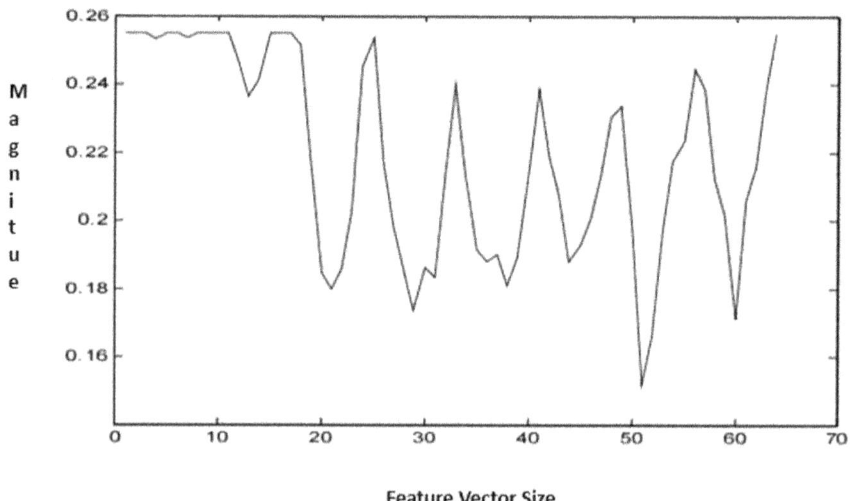

FIGURE 7.7   Concatenated 128 feature vector of face and fingerprint.

modalities and 16 features from second modalities. Concatenated features of both modalities give a total of 32 feature vector.

Figure 7.8 shows how image is divided into 16 blocks, which are formed by dividing the image into 4 rows and 4 columns.

Figures 7.9 and 7.10 show 16 featur vector of face and fingerprint, respectively. Figure 7.11 shows concatenated combined feature vector of face and fingerprint,

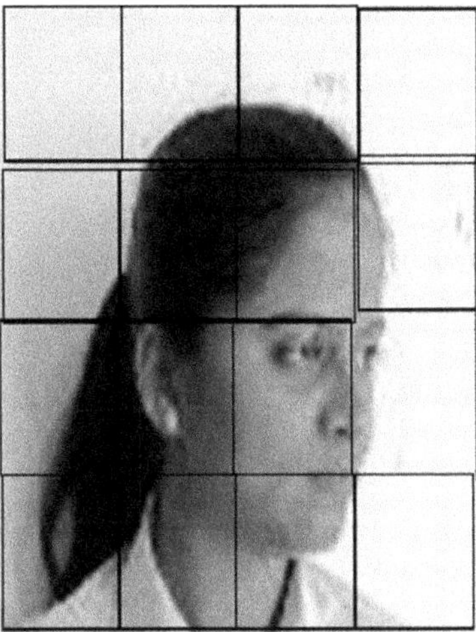

FIGURE 7.8   Dividing input image into 16 blocks.

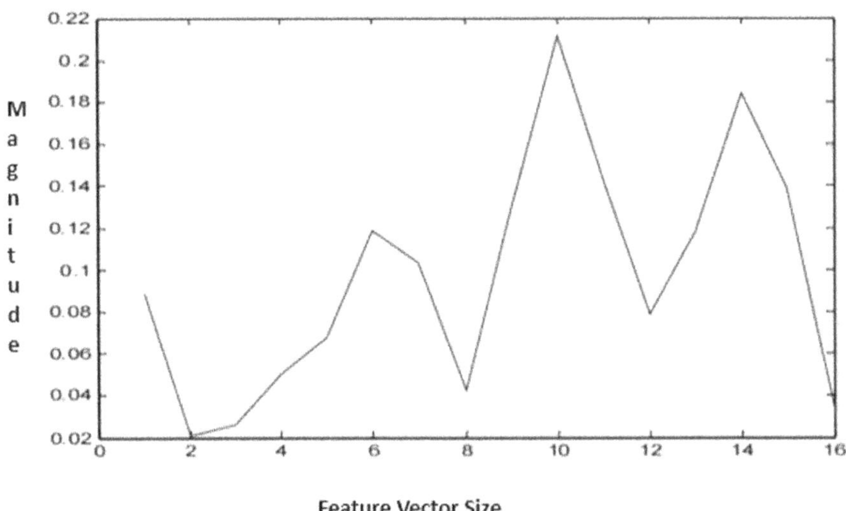

FIGURE 7.9   Face feature vector (16 features).

which shows first 16 features of face biometric and next 16 features of fingerprint biometric so a total size of feature vector after concatenation is 32. The x-axis shows the feature vector size and the y-axis shows the magnitude of features.

Results provided in Tables 7.1 and 7.2 show a comparison of feature-level fusion of face, palmprint, and fingerprint with concatenated features on the CASIA database and local database. Results show that the feature-level fusion of palmprint and

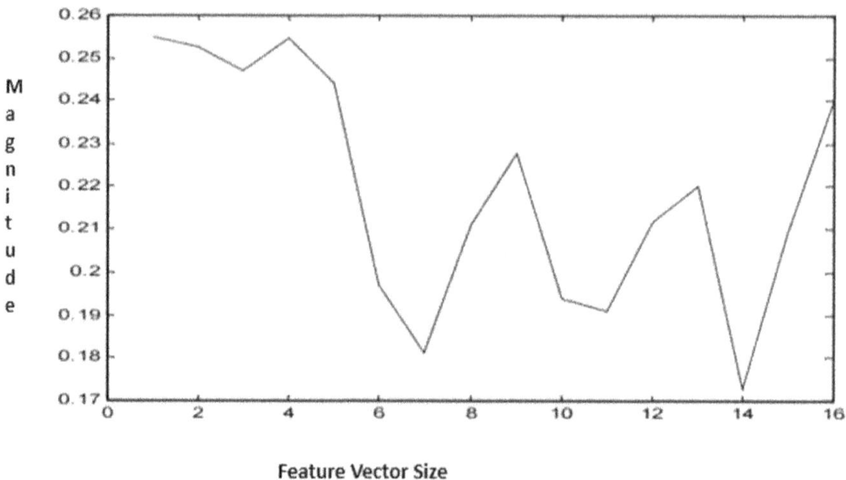

FIGURE 7.10   Fingerprint feature vector (16 features).

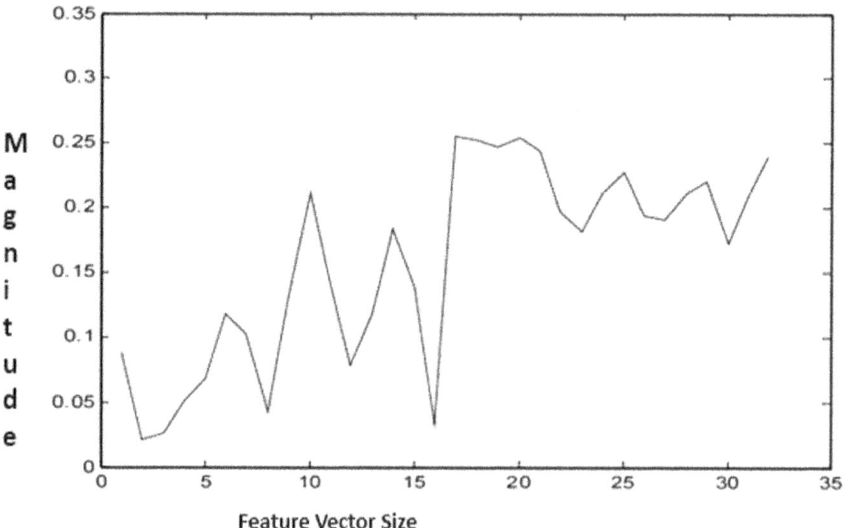

FIGURE 7.11   Concatenated 32 feature vector of face and fingerprint.

fingerprint with 128 concatenated features gives more accuracy as compared to other features.

Result provided in Table 7.3 shows a comparison of feature-level fusion of face, palmprint, and fingerprint with Sum features on the CASIA database. Results show that the feature-level fusion of palmprint and fingerprint with 64 Sum features gives more accuracy as compared to other features.

TABLE 7.1

Comparison of different biometric modality using feature-level fusion with concatenated features for the CASIA database

Biometric modality	Features	FAR	FRR	GAR	EER
Face_fingerprint	(16+16=32)	0.34	0	100	0.27
Face_Fingerprint	(64+64=128)	0.08	0	100	0.18
Face_palmprint	(16+16=32)	0.72	0.02	98	0.282
Face_palmprint	(64+64=128)	0.41	0.01	99	0.1475
Palm_fingerprint	(16+16=32)	0.8	0.01	99	0.25
Palm_fingerprint	(64+64=128)	0.7	0	100	0.11

TABLE 7.2

Comparison of different biometric modality using feature-level fusion with concatenated features for the local database

Biometric modality	Features	FAR	FRR	GAR	EER
Face_fingerprint	(16+16=32)	0.42	0.01	99	0.35
Face_Fingerprint	(64+64=128)	0.16	0	100	0.28
Face_palmprint	(16+16=32)	0.82	0.04	96	0.382
Face_palmprint	(64+64=128)	0.51	0.03	97	0.2275
Palm_fingerprint	(16+16=32)	0.84	0.02	98	0.29
Palm_fingerprint	(64+64=128)	0.8	0	100	0.19

TABLE 7.3

Comparison of different biometric modality using feature-level fusion with Sum features for the CASIA database

Biometric Modality	Features	FAR	FRR	GAR	EER
Face_fingerprint	16	0.19	0.02	98	0.435
Face_Fingerprint	64	0.16	0.027	97.3	0.335
Face_palmprint	16	0.18	0.04	96	0.212
Face_palmprint	64	0.15	0.03	97	0.201
Palm_fingerprint	16	0.12	0.01	99	0.1775
Palm_fingerprint	64	0.03	0	100	0.142

Result provided in Table 7.4 shows a comparison of feature-level fusion of face, palmprint, and fingerprint with Sum features on the local database. Results show that the feature-level fusion of palmprint and fingerprint with 64 Sum features gives more accuracy as compared to other features.

Result provided in Table 7.5 shows a comparison of feature-level fusion of face, palmprint, and fingerprint with maximum features on the CASIA database. Results

TABLE 7.4

Comparison of different biometric modality using feature-level fusion with Sum features for the local database

Biometric modality	Features	FAR	FRR	GAR	EER
Face_fingerprint	16	0.21	0.03	97	0.54
Face_Fingerprint	64	0.25	0.04	96	0.41
Face_palmprint	16	0.24	0.05	95	0.292
Face_palmprint	64	0.19	0.03	97	0.231
Palm_fingerprint	16	0.15	0.02	98	0.205
Palm_fingerprint	64	0.08	0.01	99	0.154

TABLE 7.5

Comparison of different biometric modality using feature-level fusion with Maximum features for the CASIA database

Biometric Modality	Features	FAR	FRR	GAR	EER
Face_fingerprint	16	0.58	0.08	92	0.615
Face_Fingerprint	64	0.91	0.02	98	0.505
Face_palmprint	16	0.91	0.04	96	0.45
Face_palmprint	64	0.74	0.02	98	0.41
Palm_fingerprint	16	0.61	0.01	99	0.24
Palm_fingerprint	64	0.42	0	100	0.21

TABLE 7.6

Comparison of different biometric modality using feature-level fusion with Maximum features for the local database

Biometric Modality	Features	FAR	FRR	GAR	EER
Face_fingerprint	16	0.72	0.06	94	0.685
Face_Fingerprint	64	1.42	0.04	96	0.57
Face_palmprint	16	0.85	0.05	95	0.49
Face_palmprint	64	0.82	0.03	97	0.44
Palm_fingerprint	16	0.67	0.03	97	0.38
Palm_fingerprint	64	0.532	0.02	98	0.34

show that the feature-level fusion of palmprint and fingerprint with 64 maximum features gives more accuracy as compared to other features.

Result provided in Table 7.6 shows a comparison of feature-level fusion of face, palmprint, and fingerprint with maximum features on the local database. Results show that the feature-level fusion of palmprint and fingerprint with 64 maximum features gives more accuracy as compared to other features.

TABLE 7.7

Comparison of different biometric modality using feature-level fusion with Minimum features for the CASIA database

Biometric Modality	Features	FAR	FRR	GAR	EER
Face_fingerprint	16	0.62	0.34	66	0.485
Face_Fingerprint	64	0.76	0.07	93	0.39
Face_palmprint	16	0.57	0.06	94	0.43
Face_palmprint	64	0.77	0.04	96	0.32
Palm_fingerprint	16	0.13	0.05	95	0.22
Palm_fingerprint	64	0.025	0.03	97	0.12

TABLE 7.8

Comparison of different biometric modality using feature-level fusion with Minimum features for the local database

Biometric modality	Features	FAR	FRR	GAR	EER
Face fingerprint	16	0.70	0.21	79	0.42
Face_Fingerprint	64	0.81	0.10	90	0.35
Face_palmprint	16	0.54	0.08	92	0.56
Face_palmprint	64	0.69	0.03	97	0.41
Palm_fingerprint	16	0.25	0.06	94	0.31
Palm_fingerprint	64	0.15	0.04	96	0.18

Result provided in Table 7.7 shows a comparison of feature-level fusion of face, palmprint, and fingerprint with minimum features on the CASIA database. Results show that the feature-level fusion of palmprint and fingerprint with 64 minimum features gives more accuracy as compared to other features.

Result provided in Table 7.8 shows a comparison of feature-level fusion of face, palmprint, and fingerprint with minimum features on the local database. Results show that the feature-level fusion of palmprint and face with 64 minimum features gives more accuracy as compared to other features.

Results provided in Tables 7.1–7.8 show a comparison of feature-level fusion with concatenated features, Sum features, Maximum features, Minimum features on the CASIA database and local database. Results show that the feature-level fusion with 64 concatenated features, which means 128 features after fusion, gives more accuracy as compare to other features.

7.2.3 Concatenated Features

Let $F1$ and $F2$ be two feature vectors extracted by applying contourlet transform on random selected any two biometric modalities. The dimensions of feature vector are Pn and Sn. Two feature vectors of biometric modalities are given by

$$F1 = [P1, P2, ..., Pn]$$
$$F2 = [S1, S2, ..., Sn]$$

The new feature vector F is formed by concatenation of two feature vectors.

$$F = [F1, F2] \tag{7.1}$$

The dimension of F is equal to $Pn + Sn$. In the presented algorithm, 64 features are extracted from each of trait and concatenated to form new feature vector whose dimension $64+64 = 128$. F is stored as new template in the database for matching.

7.2.4 SUM FEATURES

The concatenated feature vector size is very high which is reduced by applying Sum features of randomly selected two biometrics, which is a very simple method. Sum of two feature vectors is given as

$$Fs = F1 + F2 \tag{7.2}$$

The dimension of combined feature vector is 128 for 64 features of individual's biometric modality and feature vector is 32 for 16 features of individual's biometric modality.

7.2.5 MAXIMUM FEATURES

Maximum value between two randomly selected biometrics is considered as a feature vector. The dimension of combined feature vector is 64 for 64 features of individual's biometric modality and feature vector is 16 for 16 features of individual's biometric modality.

$$Fm = Max(Fp, Fs) \tag{7.3}$$

7.2.6 MINIMUM FEATURES

Minimum value between two randomly selected biometric is considered as a feature vector. The dimension of combined feature vector is 64 for 64 features of individual's biometric modality and feature vector is 16 for 16 features of individual's biometric modality.

$$Fm = Min(Fp, Fs) \tag{7.4}$$

7.3 FEATURE-LEVEL FUSION USING CONTOURLET TRANSFORM FEATURES

Here the first feature is extracted using the contourlet transform and then, for feature-level fusion concatenation, Sum, maximum, and minimum of features are used to form a combined feature vector. For dimension reduction of feature vector, a linear discriminate analysis is used.

Tables 7.9 and 7.10 show the results of feature-level fusion with contourlet transform features for face and palmprint biometrics for the local database and the CASIA database. The EER value of the CASIA database for Sum feature is less as compared to other features. The EER value of the local database for minimum features is less as compared to other features.

Tables 7.11 and 7.12 show the results of feature-level fusion with contourlet transform features for face and fingerprint print biometrics for the local database and the

TABLE 7.9

Feature-level fusion with contourlet transform features for face and palmprint for the CASIA database

Biometric modality	Features	FAR	FRR	GAR	EER
Face_Palmprint	concatenated	0.22	0.01	99	0.055
Face_Palmprint	Sum	0.11	0	100	0.024
Face_palmprint	Maximum	0.38	0	100	0.059
Face_Palmprint	Minimum	0.28	0.02	98	0.04

TABLE 7.10

Feature-level fusion with contourlet transform features for face and palmprint for the local database

Biometric modality	Features	FAR	FRR	GAR	EER
Face_Palmprint	Concatenated	0.28	0	100	0.059
Face_Palmprint	Sum	0.23	0	100	0.031
Face_palmprint	Maximum	0.29	0.03	97	0.062
Face_Palmprint	Minimum	0.41	0.02	98	0.03

TABLE 7.11

Feature-level fusion with contourlet transform features for face and fingerprint for the CASIA database

Biometric Modality	Features	FAR	FRR	GAR	EER
Face_fingerprint	Concatenated	0.07	0	100	0.07
Face_Fingerprint	Sum	0.06	0	100	0.03
Face_fingerprint	Maximum	0.09	0.01	99	0.09
Face_Fingerprint	Minimum	0.1	0	100	0.08

CASIA database. The EER value of the CASIA database and the local database for Sum feature is less as compared to other features.

Tables 7.13 and 7.14 show the results of feature-level fusion with contourlet transform features for fingerprint and palmprint biometric for the local database and the CASIA database. The EER value of the CASIA database for concatenated feature is less as compared to other features. The EER value of the local database for concatenated features is less as compared to other features.

Results provided in Tables 7.9–7.14 show the results of feature-level fusion using contourlet transform features on the CASIA database and the local database. Results show the fusion using Sum features gives better recognition performance. Minimum

TABLE 7.12

Feature-level fusion with contourlet transform features for face and fingerprint for local database

Biometric modality	Features	FAR	FRR	GAR	EER
Face_fingerprint	Concatenated	0.06	0	100	0.08
Face_Fingerprint	Sum	0.07	0	100	0.05
Face_fingerprint	Maximum	0.08	0.01	99	0.09
Face_Fingerprint	Minimum	0.2	0.01	99	0.07

TABLE 7.13

Feature-level fusion with contourlet transform features for fingerprint and palmprint for the CASIA database

Biometric modality	Features	FAR	FRR	GAR	EER
Fingerprint_Palmprint	Concatenated	0.012	0	100	0.01
Fingerprint_Palmprint	Sum	0.06	0	100	0.05
Fingerprint_palmprint	Maximum	0.10	0.01	99	0.06
Fingerprint_Palmprint	Minimum	0.21	0.01	99	0.04

TABLE 7.14

Feature-level fusion with contourlet transform features for fingerprint and palmprint for the local database

Biometric modality	Features	FAR	FRR	GAR	EER
Fingerprint_Palmprint	Concatenated	0.05	0	100	0.02
Fingerprint_Palmprint	Sum	0.07	0.01	99	0.06
Fingerprint_palmprint	Maximum	0.15	0.02	98	0.07
Fingerprint_Palmprint	Minimum	0.18	0.01	99	0.05

and concatenated are very simple but gives average performance. Maximum gives poorly performance. The Sum rule perform best because it is less sensitive to errors in probability distributions, and in this rule, genuine and imposter classes are not ambiguous and hence feature level strongly enhances prior probabilities.

7.4 NORMALISATION TECHNIQUE FOR HAND GEOMETRY FEATURES

The face, fingerprint, palm print, and hand modality contain the heterogeneous feature vector. To modify the location and scale parameters of hand geometry modalities, Min_Max normalising technique is used which transform the values into common domain before concatenating them to form a single one. Assumed that feature vector values of hand geometry are [0, 100].

Let x and x' are feature vector values before and after normalisation.

The min–max normalisation is given as

$$X' = \frac{x - min(Fx)}{max(Fx) - min(Fx)} \tag{7.5}$$

$$min(Fx) = 0, \quad max(Fx) = 100$$

Hand geometry features before normalisation									
26	83	22	22	31	96	21	21	29	84

Hand geometry features after normalisation									
0.26	0.83	0.22	0.22	0.31	0.96	0.21	0.21	0.29	0.84

Figure 7.12 shows concatenation of feature vector after applying normalisation techniques for hand geometry features to transform the values into common domain.

Figures 7.13 and 7.14 show concatenated features of hand geometry and fingerprint with and without applying normalisation for hand geometry features. The x-axis gives total features after concatenation and the y-axis gives the amplitude of features.

Figures 7.15 and 7.16 show concatenated features of hand geometry and palmprint with and without applying normalisation for hand geometry features. The x-axis gives total features after concatenation and the y-axis gives the amplitude of features.

Figures 7.17 and 7.18 show the concatenated features of hand geometry and face with and without applying normalisation for hand geometry features. The x-axis gives total features after concatenation and the y-axis gives the amplitude of features.

Result provided in Table 7.15 shows the feature-level fusion of hand geometry features with palmprint, face, and fingerprint using concatenation features for the CASIA database. Result shows the fusion of fingerprint and hand geometry gives more accuracy compare to palmprint and face.

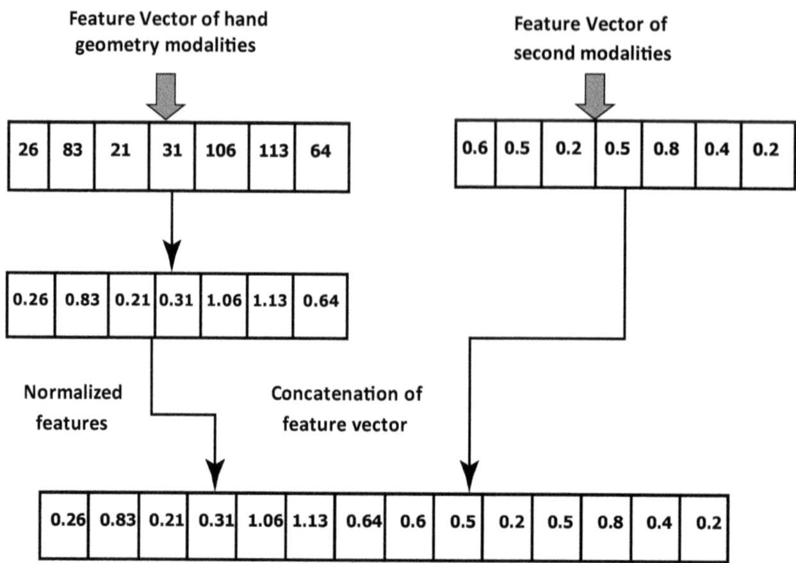

FIGURE 7.12   Feature vector after normalisation.

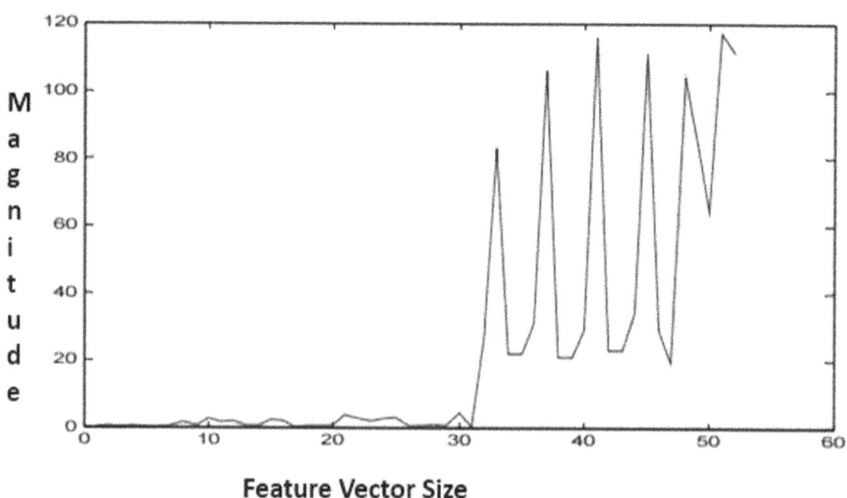

FIGURE 7.13   Hand geometry and fingerprint concatenated features without normalisation.

Result provided in Table 7.16 shows the feature-level fusion of hand geometry features with palmprint, face, and fingerprint using concatenation features for the local database. Result shows the EER value of fusion of fingerprint and hand geometry is less as compared to fusion of hand geometry with face and palmprint.

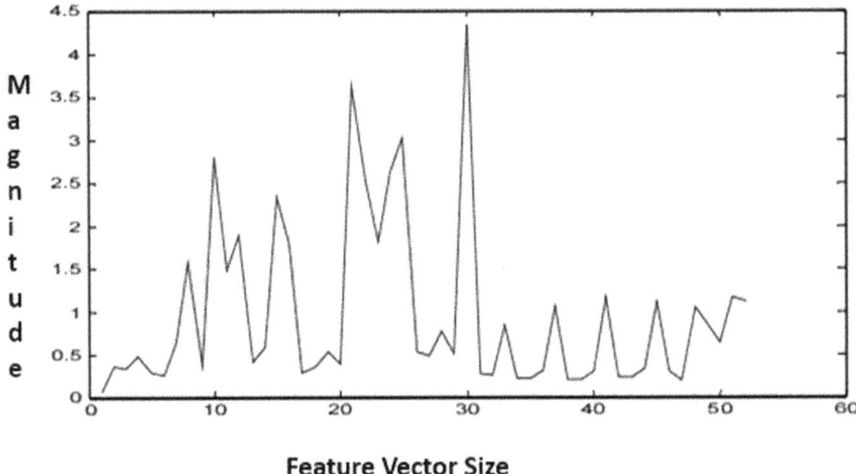

FIGURE 7.14 Hand geometry and fingerprint concatenated features with normalisation.

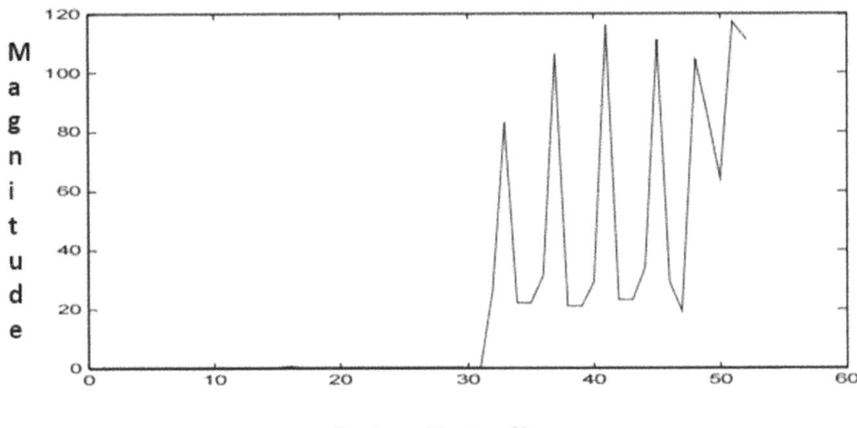

FIGURE 7.15 Hand geometry and Palmprint concatenated features without normalisation.

7.5 LINEAR DISCRIMINATE ANALYSIS (LDA)

LDA produces combinations of independent features that maximise the mean variance of the expected classes. The main idea of LDA is to find the separation of different clusters of a particular group after transformation, which can be done by matrix analysis. The goal of LDA is to maximise the between-class scatter matrix metric while minimising the within-class scatter matrix metric.

The basic steps in LDA are as follows:

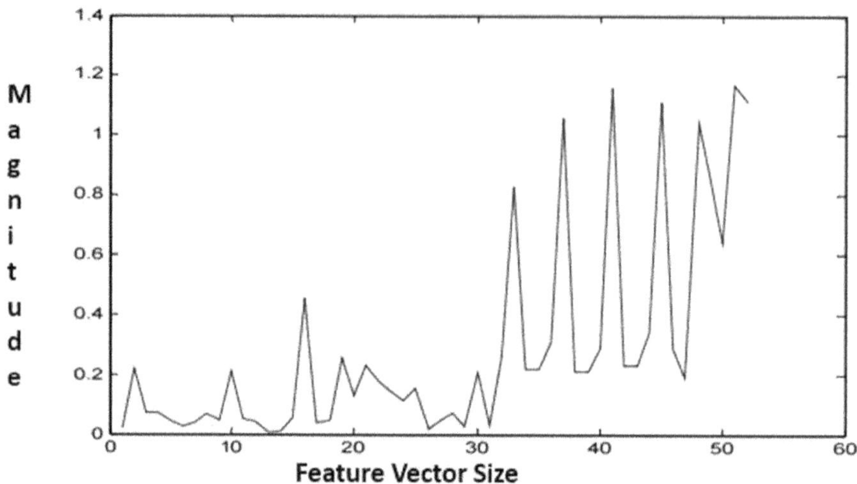

FIGURE 7.16 Hand geometry and palmprint concatenated features with normalisation.

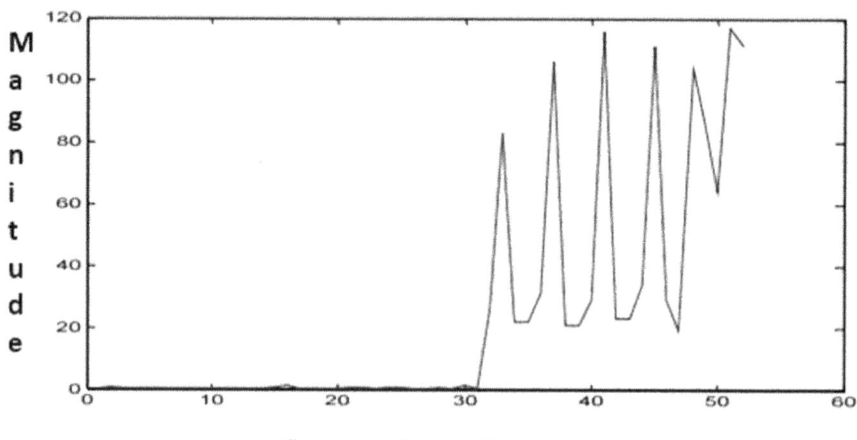

FIGURE 7.17 Hand geometry and face concatenated features without normalisation.

- Calculate the within-class scatter matrix, S_w

$$S_W = \sum_{i=1}^{NJ} \sum_{j=1}^{C} (x_i^j - \mu_j)(x_i^j - \mu_j)^T$$

where x_i^j is the ith sample pf class j,

  Mj is the mean of class j,

  C is the number of classes,

  Nj is the number of samples in class j.

  Calculate the between-class scatter matrix, Sb

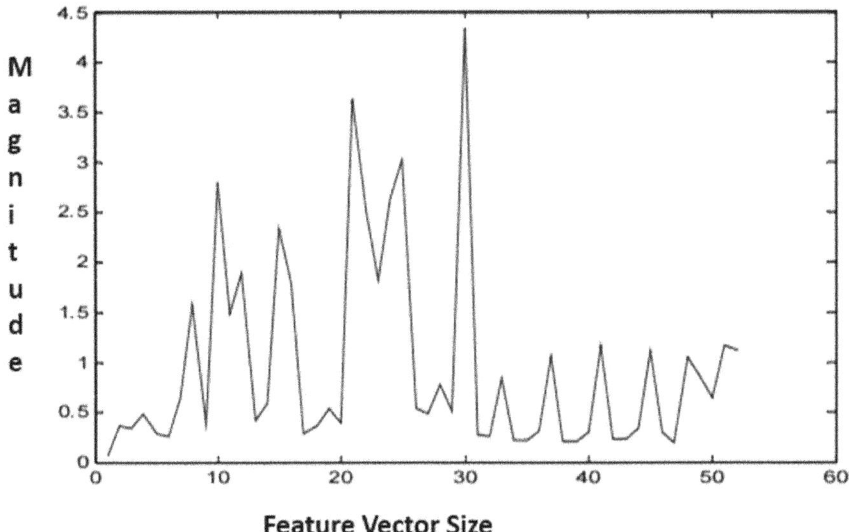

FIGURE 7.18   Hand geometry and face concatenated features with normalisation.

TABLE 7.15

Feature-level fusion of hand geometry features with concatenation of features for the CASIA database

Biometric Modality	FAR	FRR	GAR	EER
Palmprint_Hand geometry	0.23	0.015	98.5	0.120
Face_Hand geometry	0.312	0.02	98	0.231
Fingerprint_Hand geometry	0.40	0.008	99.2	0.052

TABLE 7.16

Feature-level fusion of hand geometry features with concatenation of features for the local database

Biometric Modality	FAR	FRR	GAR	EER
Palmprint_Hand geometry	0.94	0.015	97.5	0.15
Face_Hand geometry	0.88	0.04	96	0.30
Fingerprint_Hand geometry	0.85	0.015	98.5	0.068

TABLE 7.17

Feature vector size before and after applying the LDA algorithm

Biometric modality	Feature vector size before applying LDA	Feature vector size after applying LDA
Fingerprint_Palmprint	62	20
Fingerprint_Face	62	20
Palmprint_Face	62	20
Fingerprint_Hand geometry	52	20
Palmprint_Handgeometry	52	20
Face_Handgeometry	52	20

$$S_b = \sum_{i=1}^{C} (\mu_j - \mu)(\mu_j - \mu)^T$$

where μ represents the mean of all classes.

Calculate the eigenvectors of the projection matrix

$$W = eig(S_W^{-1} - S_b)$$

Compare the test image's projection matrix with the projection matrix of each training image by using a similarity measure. The result is the training image which is the closest to the test image.

Table 7.17 shows how the feature-level fusion reduces the feature vector size after using the LDA algorithm. After using the contourlet instead, a total of 31 features for the fingers, palm, and face emerge. Therefore, after feature concatenation of feature-level fusion, the total feature vector becomes 62 (31 + 31). The feature vector size becomes 20 after applying LDA.

The total number of features for hand geometry is 21. Therefore, after feature fusion of feature-level fusion, the total feature vector becomes 52 (31 + 21). The feature vector size becomes 20 after applying LDA.

7.6   SUMMARY

This chapter discusses feature-level fusion of two randomly selected biometric. For feature-level fusion, block variance features and contourlet transform features are calculated. The face, fingerprint, palm print, and hand modality contains the heterogeneous feature vector so the normalisation process is used to change the position and measure the geometric modalities of the hand.

ROC curves and EER values validate the experiments performed for sensor-level and feature-level fusion. For fusing two randomly selected modalities, feature-level fusion with contourlet transform obtained the best results as compared to block variance features as contourlet transform decompose the image into low-frequency and high-frequency contour coefficient with different scales and various angles. For dimension, the reduction of feature vector LDA is used.

8 Result and Discussion

8.1 RESULT AND DISCUSSION

Tests were conducted to determine the accuracy of the presented method. This chapter discusses a novel method of multimodal biometric techniques that uses a random selection of biometrics at sensor-level and feature-level fusion. Receiver operating characteristics (ROC) curve and equal error rate (EER) values prove the tests of unimodal and multimode biometric systems. This chapter also introduces the design of a multimodal biometric identification system. It begins with a brief discussion and summary of the results from each chapter.

8.1.1 DATABASES USED

The experimental analysis presented in this book is based on the standard database (CASIA database) and local data, from which we collected data from 200 different individuals and five samples of each individual person, i.e., a total of 1,000 images per modalities. Hand and fingerprint information will be subject to change due to in illumination, ageing, and line orientation. Changes in illumination, ageing, expression, posing style, and angle occur in facial images.

8.1.2 RESULTS OF PERFORMANCE MEASUREMENT PARAMETERS OF THE BIOMETRIC SYSTEMS

A biometric identification system provides a match score by comparing feature template of enrolment biometric data with feature vector of identity claim person. If both feature vectors are close to each other, it generates a higher match score. For decision-making, one threshold value is set. A person is accepted if the match score is more than a threshold value, otherwise person is rejected. A genuine user is rejected if the threshold value is very high and an imposter user is accepted if the threshold value is set too low. Threshold selection should be proper to get the highest accuracy as both false acceptance rate and false rejection rate depend on threshold value.

Table 8.1 shows different threshold false accept rate (FAR), false reject rate (FRR), and genuine acceptance rate (GAR) values for unimodal fingerprint recognition using energy features. After threshold 0.5, FAR, FRR, and GAR values become constant. FAR, FRR, and GAR values are calculated by using TP, TN, FP, and FN values.

Figure 8.1 shows FAR and FRR values for various threshold unimodal fingerprint identification systems using energy features. The EER (equal error rate) is a point where FAR and FRR coincide with each other. EER is present in between 0.2 and 0.3

DOI: 10.1201/9781032665993-8

TABLE 8.1

Threshold values and corresponding FAR, FRR, and GAR values

Sr. No	Threshold	TP	TN	FP	FN	FAR	FRR	GAR
1	0	0	100	0	100	0	1	0
2	0.1	19	100	0	81	0	0.81	19
3	0.15	62	99	1	38	0.01	0.38	62
4	0.2	81	94	6	19	0.06	0.19	81
5	**0.25**	**95**	**54**	**46**	**5**	**0.46**	**0.05**	**95**
6	0.3	99	30	70	1	0.7	0.01	99
7	0.4	100	21	79	0	0.79	0	100
8	0.5	100	3	97	0	0.97	0	100
9	0.55	100	0	100	0	1	0	100
10	0.6	100	0	100	0	1	0	100
11	0.7	100	0	100	0	1	0	100
12	0.8	100	0	100	0	1	0	100
13	0.9	100	0	100	0	1	0	100
14	1	100	0	100	0	1	0	100

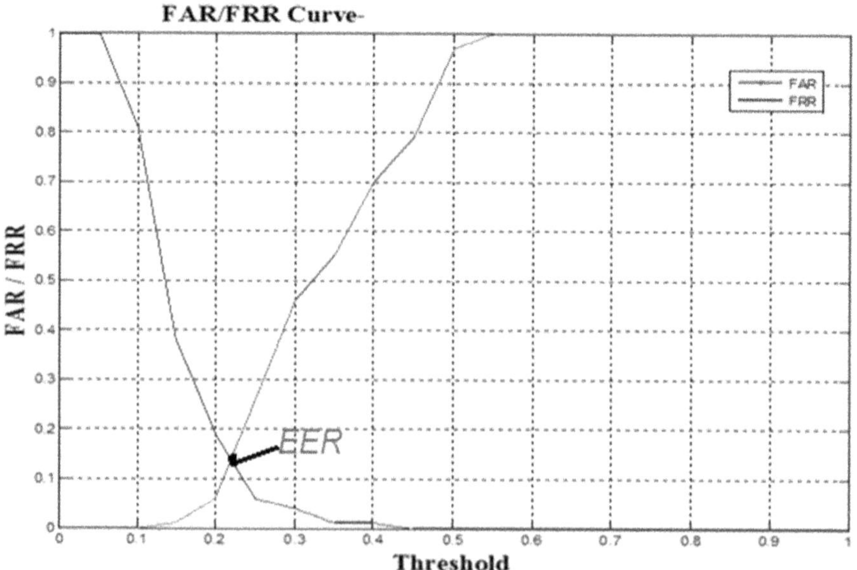

FIGURE 8.1   Graph of threshold vs FAR/FRR.

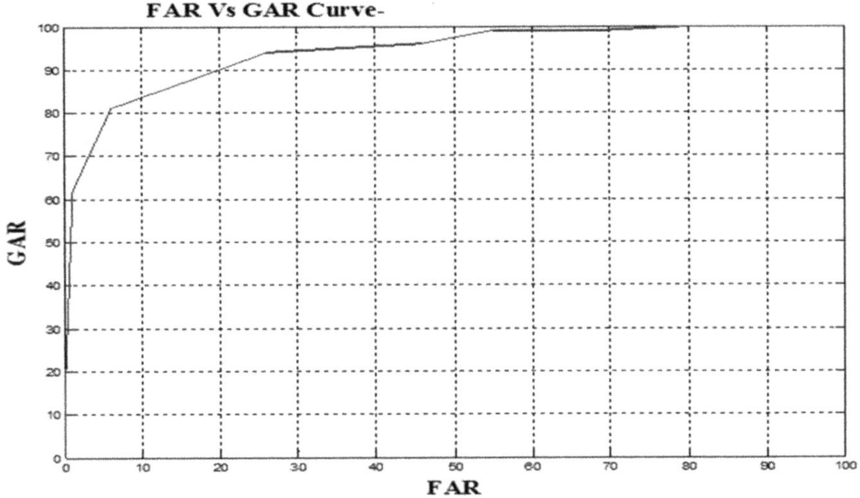

FIGURE 8.2 ROC curve (FAR vs GAR).

thresholds. From Table 4.1, the threshold is 0.25 so the corresponding FAR = 0.46, FRR = 0.05, and GAR = 95. The same procedure is applied to get FAR, FRR, and GAR values of all the algorithms.Receiver operating characteristics is plot of GAR against each value of FAR for various threshold as shown in Figure 8.2

For threshold 0.15

$$FAR = \frac{1}{1+99} = 0.01$$

$$FRR = \frac{38}{38+62} = 0.38$$

$$GAR = 1 - 0.38 = 62\%$$

For threshold 0.25

$$FAR = \frac{46}{46+54} = 0.46$$

$$FRR = \frac{5}{5+95} = 0.05$$

$$GAR = 1 - 0.04 = 95\%$$

For threshold 0.6

$$FAR = \frac{100}{100+0} = 1$$

$$GAR = 1-0 = 100\%$$

For threshold 0.15, the FAR value is 0.01, the FRR value is 0.38, and the GAR value is 62%. For threshold 0.25, the FAR value is 0.46, the FRR value is 0.05, and the GAR value is 95%. For threshold 0.6, the FAR value is 1, the FRR value is 0, and the GAR value is 100%. The same processes are applied for getting FAR, FRR, and GAR and EER values for different techniques.

8.1.3 RESULTS OF PERFORMANCE MEASUREMENT PARAMETERS OF MULTIMODAL RECOGNITION SYSTEM

Unimodal biometric systems face many challenges such as noise in sensed data, intra-class variations, inter-class similarities, and non-universal and spoof attacks multimodal biometric fusion is used to overcome the disadvantages of unimodal systems and increase the accuracy of the system. This book presents sensor-level and feature-level fusion using randomly selected biometric features. For dimension reduction feature vector size, linear discriminate analysis (LDA) is used after contourlet transform feature-level fusion.

Table 8.2 shows a comparison of unimodal and multimodal biometric systems that use face and fingerprint modalities. Feature-level fusion with contourlet transform gives better performance than feature-level block features, sensor-level fusion, and its unimodal counterparts.

Figure 8.3 shows a comparison of GAR values between unimodal and multimodal biometric systems using face and fingerprint modalities. The results show that the GAR value of feature-level fusion using contourlet transform is more than the GAR values of feature-level fusion using block features, sensor-level fusion, and its unimodal system.

TABLE 8.2

Comparison of fingerprint and face multimodal identification system

Biometric modality	FAR	FRR	GAR	EER
Unimodal face	0.71	0.088	91.2	0.355
Unimodal fingerprint	0.786	0.02	98	0.1016
Sensor-level fusion	0.3563	0	100	0.107
Feature-level fusion with block features	0.15	0.01	99	0.113
Feature-level fusion with contourlet transform features	0.06	0	100	0.03

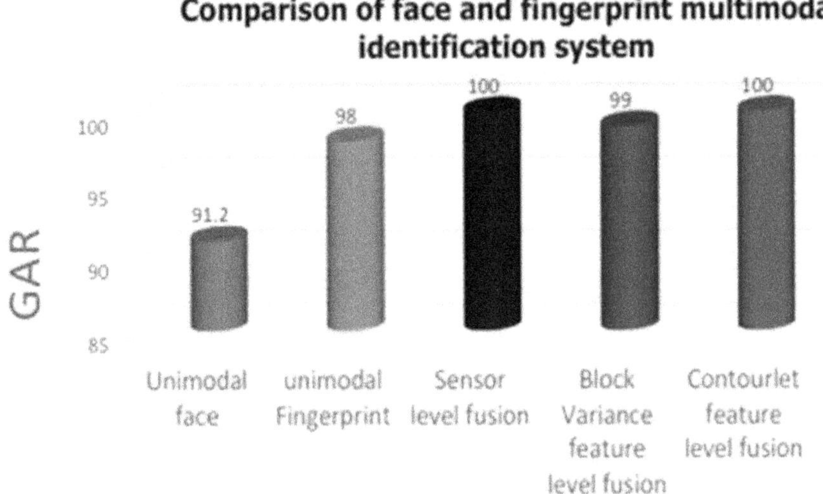

FIGURE 8.3 Comparison of GAR values between unimodal and multimodal biometric systems using face and fingerprint modalities.

Figure 8.4 is a graph of the ROC curve (FAR vs GAR), which shows a comparison between unimodal and multimodal biometric systems using face and fingerprint modalities. It shows that the GAR value of feature-level fusion using contourlet transform is more than the GAR values of feature-level fusion using block features, sensor-level fusion, and its unimodal system. The performance of feature-level fusion using contourlet transform is better as compared to other techniques. The GAR value of the unimodal face identification system is 91.2 and the fingerprint identification system is 98.

The GAR value for feature-level fusion using block features is 99. GAR values for sensor-level fusion and feature-level fusion using contourlet transform are 100. But the EER value of contourlet features is less as compared to sensor-level fusion. So the performance of feature-level fusion using contourlet transform is better than sensor-level fusion.

Table 8.3 shows a comparison between unimodal and multimodal biometric systems using face and palmprint biometric modalities. Feature-level fusion with contourlet transform gives better performance to the feature-level block features, sensor-level fusion, and its unimodal counterparts. The EER value of the unimodal face identification system is 0.355 and the palmprint identification system is 0.2175. The EER value for feature-level fusion using block features is 0.1475. The EER value for sensor-level fusion is 0.04 and the minimum EER value for feature-level fusion using contourlet transform is 0.024.

Figure 8.5 shows a comparison of GAR values of unimodal and multimodal biometric systems using face and palmprint modalities. The results show that the GAR value of the fusion technique using contourlet transform is larger than that of the

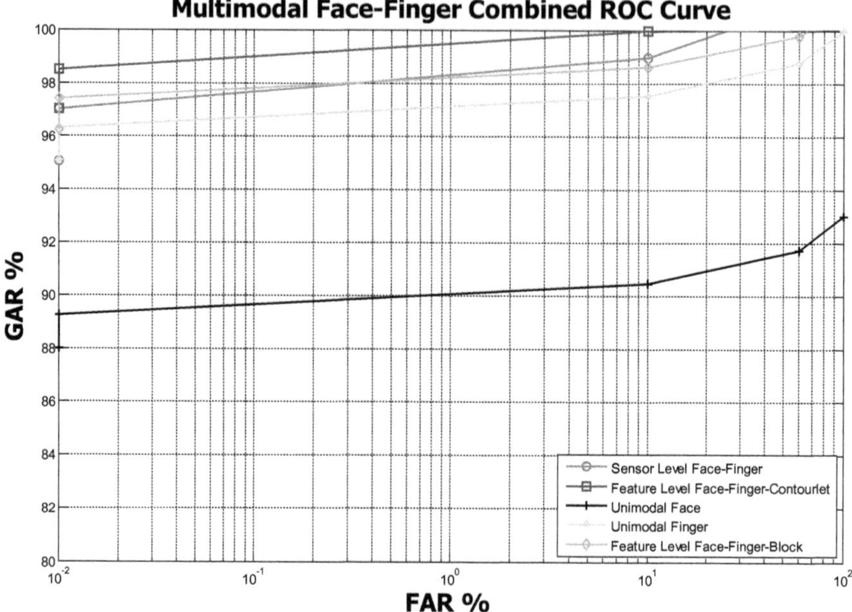

FIGURE 8.4   Multimodal face and fingerprint combined ROC curve.

TABLE 8.3

Comparison of palmprint and face multimodal identification system

Biometric modality	FAR	FRR	GAR	EER
Unimodal face	0.71	0.088	91.2	0.355
Unimodal palmprint	0.754	0.045	95.5	0.2175
Sensor-level fusion	0.312	0	100	0.04
Feature-level fusion with block features	0.41	0.03	98	0.1475
Feature-level fusion with contourlet transform features	0.11	0	100	0.024

fusion technique using feature-level block features, sensor-level fusion, and their unimodal systems.

Figure 8.6 is a graph of ROC curve (FAR vs GAR), which shows a comparison between unimodal and multimodal biometric systems using face and palmprint modalities. It shows the GAR value of feature-level fusion using contourlet transform is more than the GAR values of feature-level fusion using block features, sensor-level fusion, and its unimodal system. The performance of feature-level fusion using contourlet transform is better as compared to other techniques. The GAR value of the

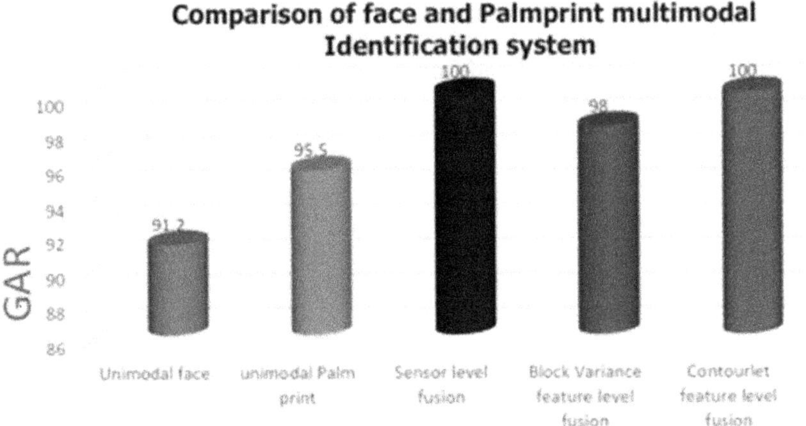

FIGURE 8.5 Comparison of GAR values between unimodal and multimodal biometric systems using face and palmprint modalities.

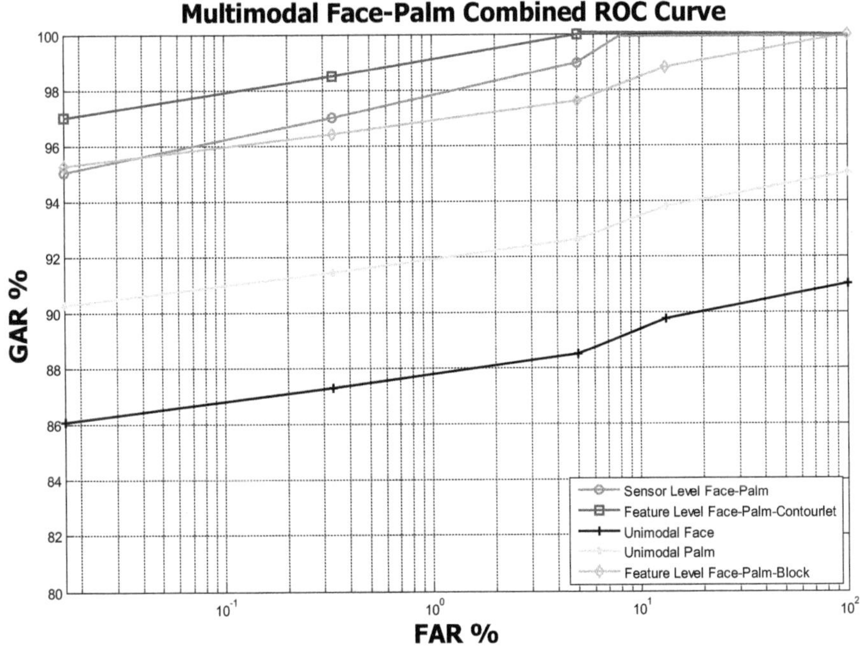

FIGURE 8.6 Multimodal face and palmprint combined ROC curve.

unimodal face identification system is 91.2 and the palmprint identification system is 95.5. The GAR value for feature-level fusion using block features is 98. GAR values for sensor-level fusion and feature-level fusion using contourlet transform are 100. But the EER value of contourlet features is less as compared to sensor-level fusion.

TABLE 8.4

Comparison of palmprint and fingerprint multimodal identification system

Biometric modality	FAR	FRR	GAR	EER
Unimodal palm	0.754	0.045	95.5	0.2175
Unimodal fingerprint	0.786	0.02	98	0.1016
Sensor-level fusion	0.257	0	100	0.0405
Feature-level fusion with block features	0.65	0	100	0.11
Feature-level fusion with contourlet transform features	0.012	0	100	0.01

So the performance of feature-level fusion using contourlet transform is better than sensor-level fusion.

Table 8.4 shows a comparison between unimodal and multimodal biometric systems using fingerprint and palmprint modalities. Feature-level fusion with contourlet transform gives better performance than feature-level block features, sensor-level fusion, and its unimodal counterparts.

Figure 8.7 shows a comparison of GAR values between unimodal and multimodal biometric systems using fingerprint and palmprint modalities. It shows the GAR values of feature-level fusion using contourlet transform, feature-level fusion using block features, and sensor-level fusion are more than its unimodal system.

Figure 8.8 is an ROC curve plot (FAR vs GAR) showing a comparison of unimodal and multimode biometric systems using fingerprint and palmprint biometric

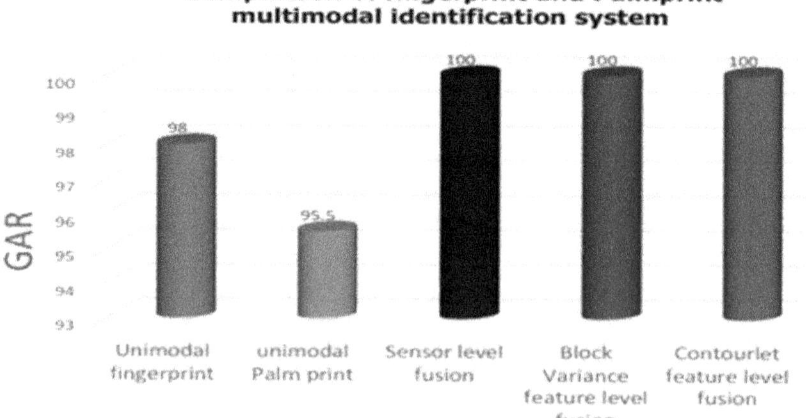

FIGURE 8.7 Comparison of GAR values between unimodal and multimodal biometric systems using fingerprint and palmprint modalities.

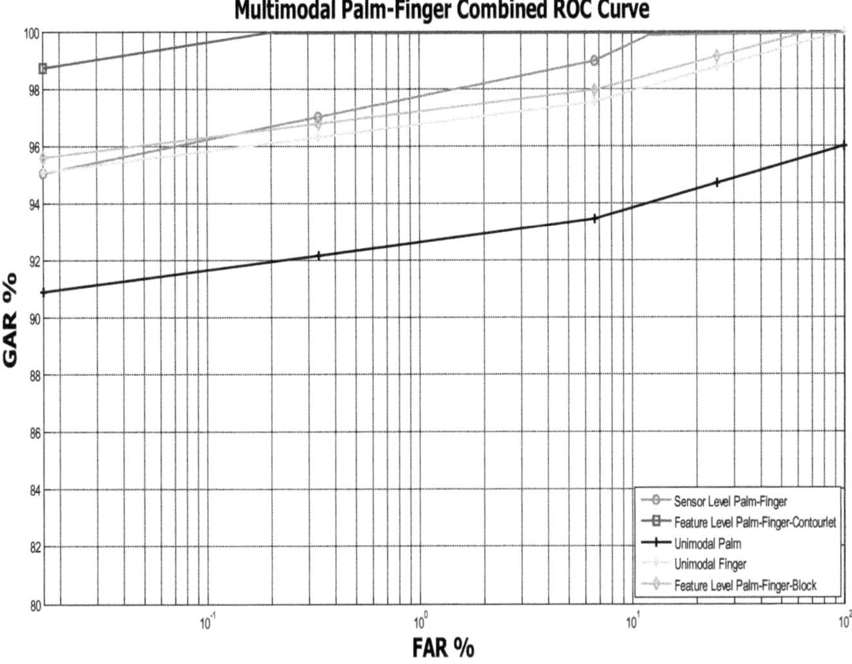

FIGURE 8.8 Multimodal face and palmprint combined ROC curve.

methods. The results show that the feature-level fusion using countourlet transform GAR value is greater than its unimodal system. Feature-level fusion using contourlet transform performs better compared to other techniques. The GAR value of the unimode fingerprint recognition is 98, and the facial recognition is 95.5. Feature-level fusion using block features, sensor-level fusion, and feature-level fusion using contourlet transform have a GAR of 100. However, the EER of contourlet features is smaller than sensor-level fusion and feature-level fusion using block features. Therefore, the performance of feature-level fusion using contourlet transform is better than sensor-level fusion.

Table 8.5 shows a comparison between unimodal and multimodal biometric systems using face and hand geometry modalities. Feature-level fusion gives better performance than its unimodal counterparts.

Figure 8.9 shows a comparison of GAR values between unimodal and multimodal biometric systems using face and hand geometry modalities. GAR value indicates that feature-level fusion is more accurate than its unimodal system.

Figure 8.10 shows a comparison of unimodal and multimodal biometric systems using face and hand geometry methods. Feature-level fusion provides better performance than its unimodal system. The GAR value of the face recognition system using the contourlet transform is 91.2. The GAR value of the hand geometry unimodal system using 21 geometric features is 90. After combining the feature vectors,

TABLE 8.5

Comparison of face and hand geometry multimodal identification system

Biometric modality	FAR	FRR	GAR	EER
Unimodal face	0.71	0.088	91.2	0.355
Unimodal hand geometry	0.85	0.10	90	0.3885
Feature-level fusion	0.312	0.02	98	0.258

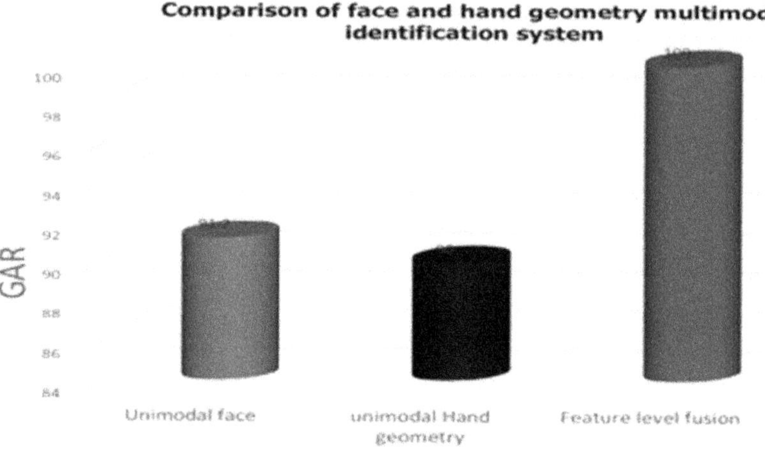

FIGURE 8.9 Comparison of GAR values between unimodal and multimodal biometric systems using fingerprint and palmprint modalities.

the multimodal system gives a GAR value of 98. Fusion at the sensor level is not possible due to heterogeneous features between face and hand geometry.

Table 8.6 shows a comparison of unimodal and multimodal biometric systems using fingerprint and hand geometry. Feature-level fusion provides better performance than its unimodal counterparts.

Figure 8.11 shows a comparison of GAR values of unimodal and multimodal biometric systems using fingerprint and hand geometry methods. The results show that the GAR of the feature-level fusion is greater than that of the unimodal system.

Figure 8.12 shows a comparison between unimodal and multimodal biometric systems using fingerprint and hand geometry biometric modalities. Feature-level fusion gives better performance than its unimodal counterparts. The GAR value of the unimodal fingerprint identification system is 98 using contourlet transform. The GAR value of the hand geometry unimodal system is 90 using 21 geometry features. After concatenation of the feature vector, the multimodal system gives a 99.2 GAR

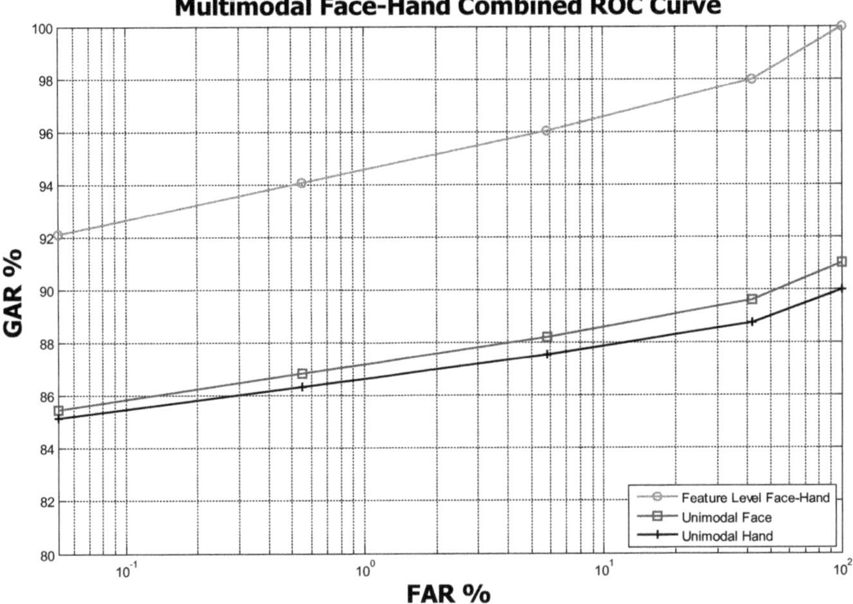

FIGURE 8.10   Multimodal face and hand geometry combined ROC curve.

TABLE 8.6
Comparison of fingerprint and hand geometry multimodal identification system

Biometric modality	FAR	FRR	GAR	EER
Unimodal fingerprint	0.186	0.02	98	0.1016
Unimodal hand geometry	0.8517	0.10	90	0.38885
Feature-level fusion	0.301	0.08	99.2	0.214

value. Sensor-level fusion is not possible due to the heterogeneous features of fingerprint and hand geometry.

Table 8.7 shows a comparison between unimodal and multimodal biometric systems using palmprint and hand geometry modalities. Feature-level fusion gives better performance than its unimodal counterparts.

Figure 8.13 shows a comparison of GAR values between unimodal and multimodal biometric systems using fingerprint and hand geometry modalities. It shows that the GAR value of feature-level fusion is more than unimodal system.

Figure 8.14 shows a comparison of unimodal and multimodal biometric systems using palmprint and hand geometry methods. Feature-level fusion provides better

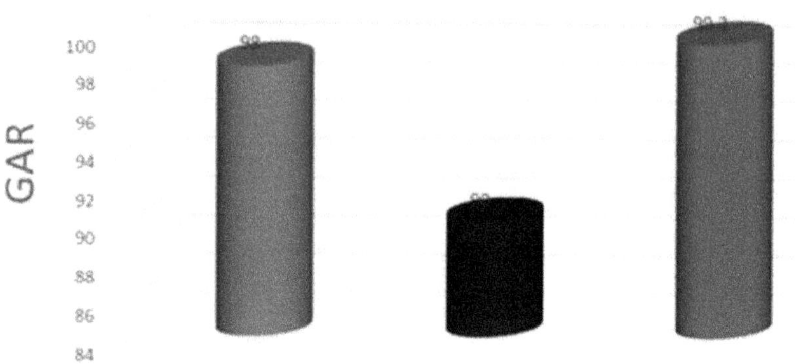

FIGURE 8.11 Comparison of GAR values between unimodal and multimodal biometric systems using fingerprint and palmprint modalities.

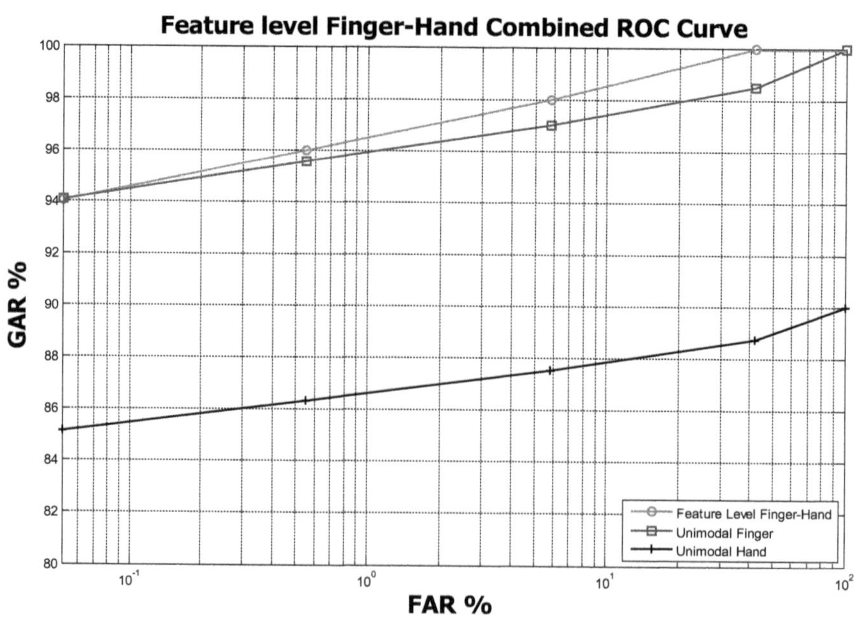

FIGURE 8.12 Multimodal fingerprint and hand geometry combined ROC curve.

TABLE 8.7

Comparison of palmprint and hand geometry multimodal identification system

Biometric modality	FAR	FRR	GAR	EER
Unimodal palmprint	0.554	0.045	95.5	0.2175
Unimodal hand geometry	0.8517	0.10	90	0.38885
Feature-level fusion	0.23	0.015	98.5	0.1845

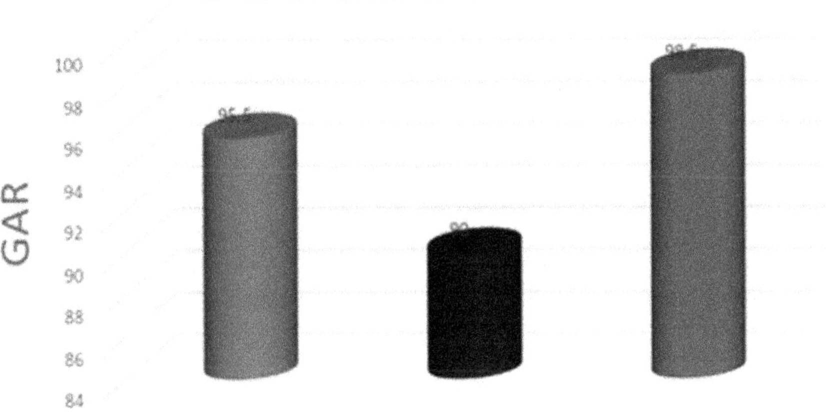

FIGURE 8.13 Comparison of GAR values between unimodal and multimodal biometric systems using fingerprint and palmprint modalities.

performance than its unimodal counterparts. The GAR value of the unimodal palmprint recognition system using the contourlet transform is 95.5. The GAR value of the hand geometry unimodal system using 21 geometric features is 90. After combining the feature vectors, the multimodal system gives a GAR value of 98.5. Sensor-level fusion is not possible due to the heterogeneous features of fingerprint and hand geometry.

8.1.4 Score Distribution of Biometric System

The accuracy of any biometric system is determined by measuring various errors. Error is measured using graphs of genuine and imposter scores depending on their frequencies and scores. In the best biometric system, there will be no overlap

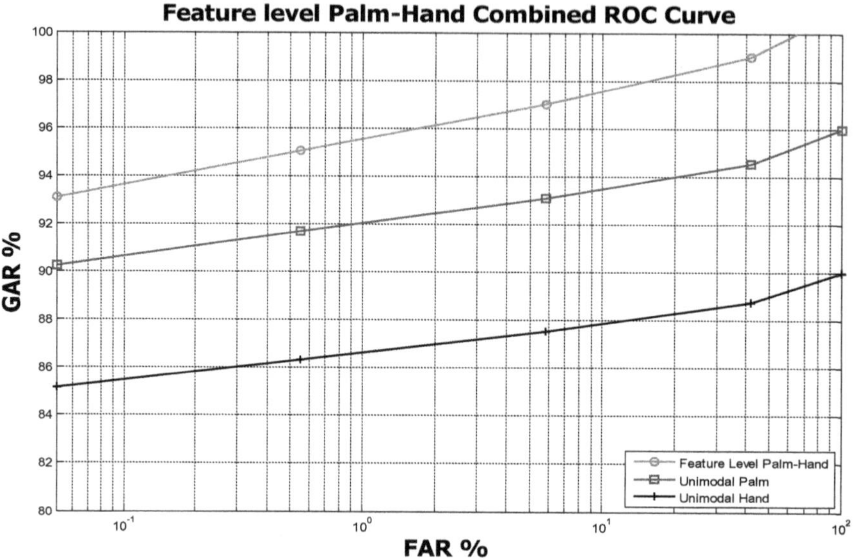

FIGURE 8.14   Multimodal palmprint and hand geometry combined ROC curve.

between genuine and imposter scores, they will not touch or overlap. So by adjusting the threshold to fit between these two distributions, easy separation occurs between genuine and imposter scores, which means the system has no false accepts or false rejects.

Since biometric systems are not 100% accurate, there is an overlap between genuine and imposter scores. Therefore, when the score falls in an overlapping area, it is not easy to determine whether it is genuine or not. The smaller the overlap, the more accurate the system. The greater the overlap, the system is less accurate. The threshold is ultimately used to determine whether the user is accepted as a real user or rejected as an unauthenticated user. An imposter score above the threshold results in a false acceptance, while a genuine score below the threshold results in a false rejection.

Figures 8.15–8.18 show genuine imposter graphs for a unimodal biometric identification system. Compared to unimodal face, palm print, and hand geometry recognition systems, unimodal fingerprint recognition has a small overlap between genuine and imposter scores. Therefore, the accuracy of the fingerprint recognition system exceeds the accuracy of face, palm, and hand geometry recognition system.

Figures 8.19–8.24 show the genuine vs imposter scores. Genuine vs imposter scores for a multimodal biometric system using feature-level fusion are higher compared to a unimodal biometric system. Compared to other biometric methods, multimodal biometric authentication using fingerprints and palmprints has less overlap between genuine vs imposter scores. Therefore, multimodal biometric identification using fingerprint and palmprint is more accurate than other multimodal biometric systems. Compared with other multimodal methods, the imposter score of hand

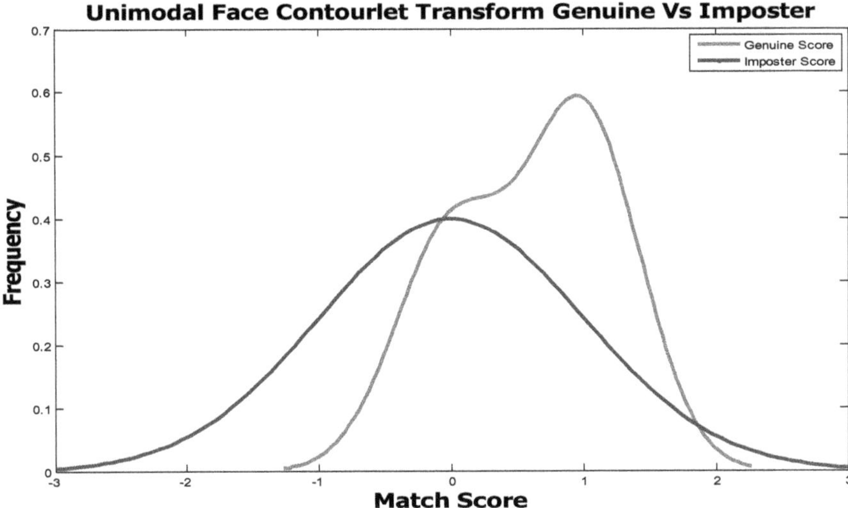

FIGURE 8.15   Genuine vs imposter score for unimodal face identification system.

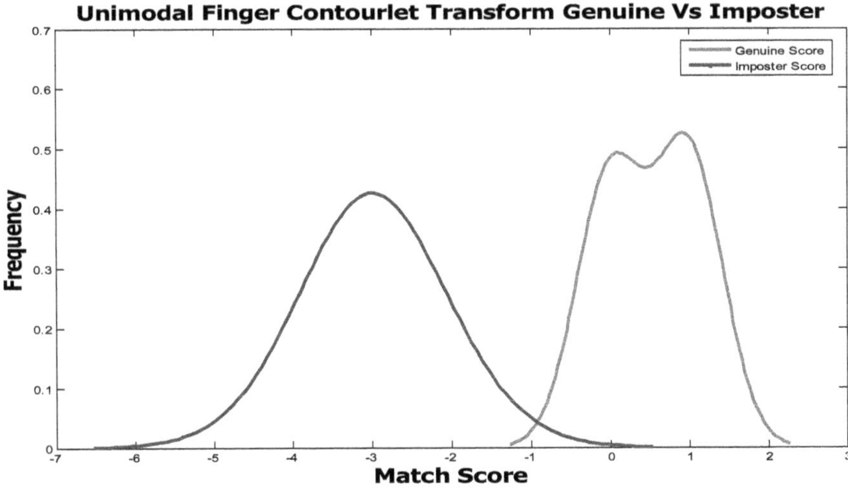

FIGURE 8.16   Genuine vs imposter score for unimodal fingerprint identification system.

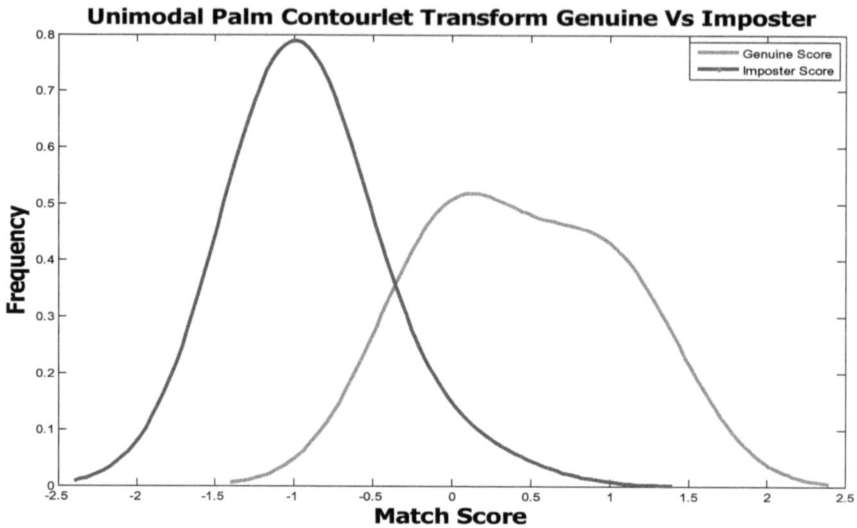

FIGURE 8.17   Genuine vs imposter score for unimodal palmprint identification system.

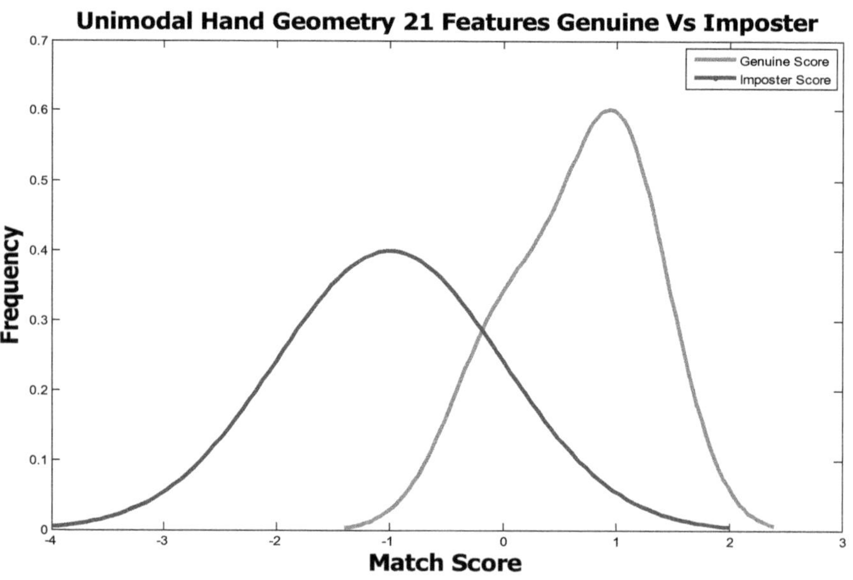

FIGURE 8.18   Genuine vs imposter score for unimodal hand geometry identification system.

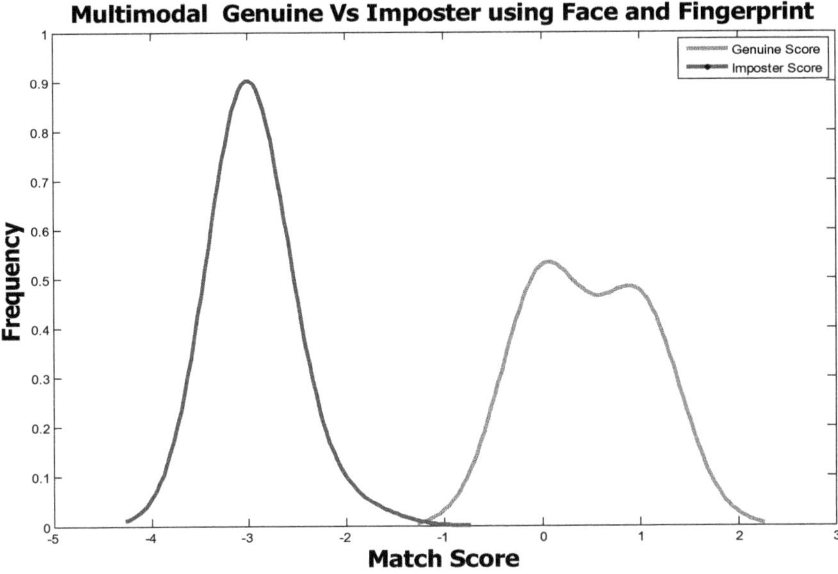

FIGURE 8.19   Genuine vs imposter score for multimodal face and fingerprint identification system.

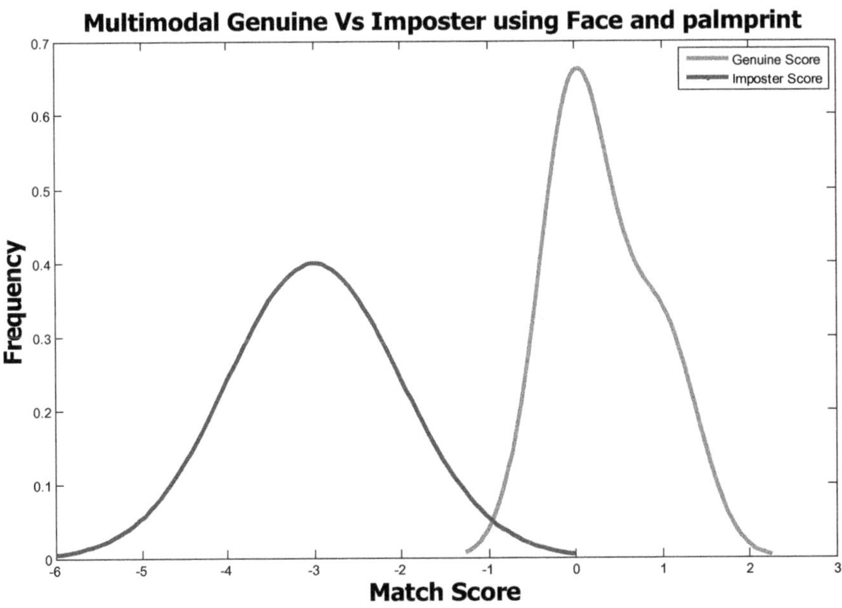

FIGURE 8.20   Genuine vs imposter score for multimodal face and palmprint identification system.

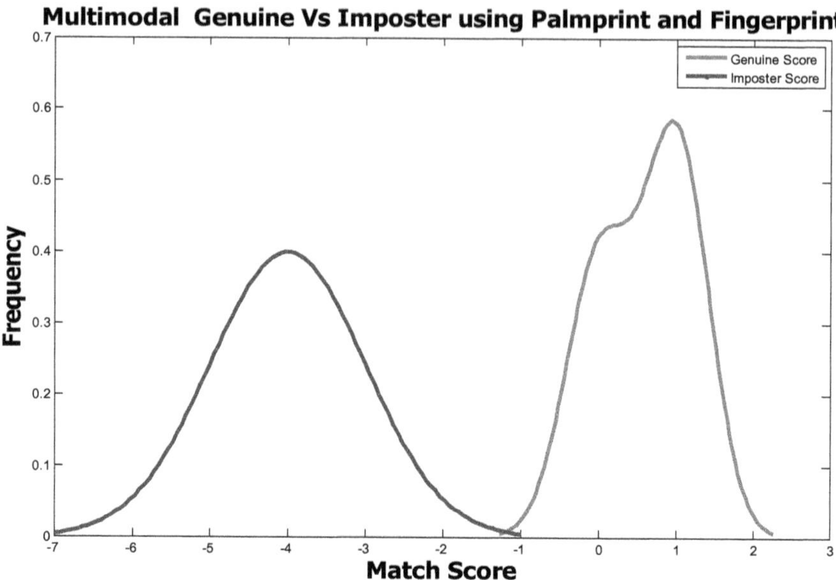

FIGURE 8.21   Genuine vs imposter score for multimodal fingerprint and palmprint identification system.

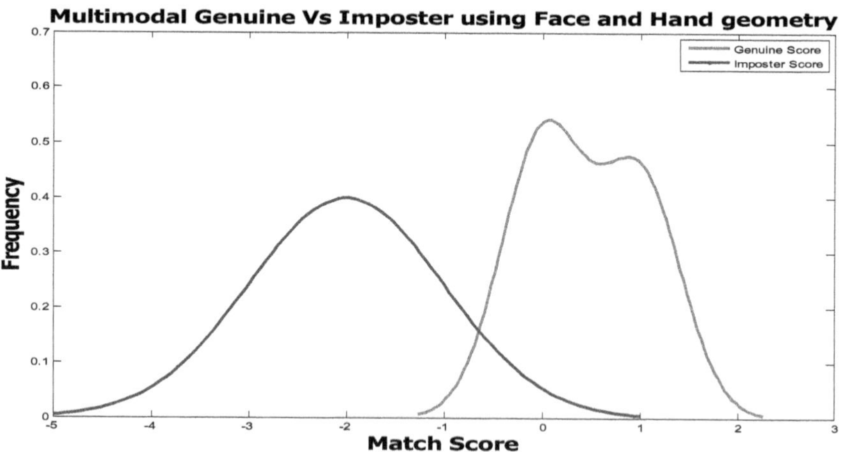

FIGURE 8.22   Genuine vs imposter score for multimodal face and hand geometry identification system.

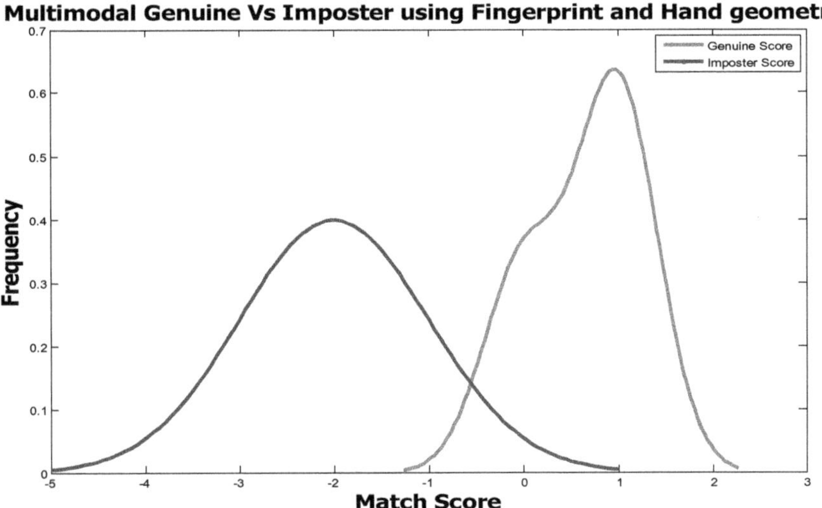

FIGURE 8.23   Genuine vs imposter score for multimodal fingerprint and hand geometry identification system.

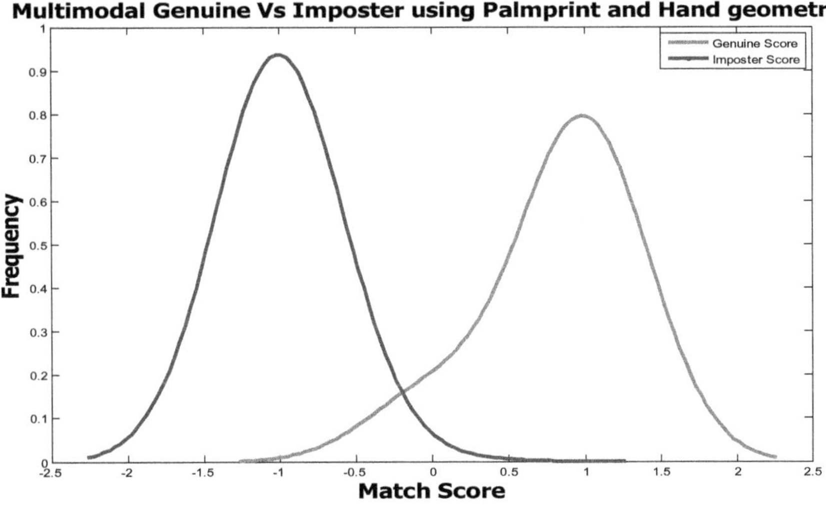

FIGURE 8.24   Genuine vs imposter score for multimodal palmprint and hand geometry identification system.

geometry and palmprint fusion is higher than the genuine ones, and the overlap area is larger, so the accuracy of hand geometry and palmprint fusion is lower as compared to other multimodal systems

8.1.5 ANALYSIS

Biometric is a characteristic measured by an individual that is discriminatory enough to be used for identification or verification purposes [2]. Popular biometric features are fingerprints, iris, retina, face, hands, and palm prints.

This book describes various feature extraction techniques for unimodal and multimodal biometric recognition systems. It presents two approaches for fingerprint recognition, minutiae based, and texture based. It also presents a texture-based approach for face and palmprint identification systems. To extract texture features, discrete wavelet transform (DWT), Gabor filter, curvelet, and contourlet transforms are used.

DWT only provides horizontal, vertical, and diagonal directional features. Wavelets do not handle curved discontinuities well. The disadvantage of the Gabor filter is that the feature vector is very large and the computation time required to extract the feature is very long, which limits the retrieval speed. The curvelet transform gives the discontinuities of the curve very well, but it converges to the continuous domain. The contourlet transformation starts from discrete domain formation and converges to continuous domain expansion. Due to its directional and anisotropic properties, the contourlet transform can represent images such as straight lines, edges, contours, and curves better than wavelet and curve transforms. The empirical study explained that contourlet transform is the best modality-specific feature extraction algorithms for fingerprint, face, and palmprint, respectively. For hand geometry, 12 features and 21 features results are compared, which shows 21 features accuracy is higher. As the number of features increases, the accuracy of the system increases.

Although biometric technology has now become an important part of personal identity management systems, the current biometric system is not 100%. Some factors that affect the accuracy of biometric systems include noise, non-universality, lack of invariant representation, and non-distinctiveness. Additionally, biometric systems are also vulnerable to security attacks. Biometric systems that combine multiple cues can overcome some of these limitations and achieve high performance, so-called multimodal biometric systems. Compared to the limitations of a single model, the problem of non-universality due to a large number of changes in face, finger, palm, and hand geometry to satisfy the population is addressed. Subjects who did not fit one modality could use another modality (e.g., a manual worker with cuts and bruises on his fingers). It is difficult for an intruder to simultaneously spoof a legitimate person's fingerprint, face, palm print, and hand geometric features.

Nowadays, researchers have shown tremendous interest in the implementation of multimodal systems using various fusion strategies.

The question is how to integrate so much data, how to determine the source of biometric data, what characteristics, features, or types of data to be integrated, and what method should be used. Biometric integration occurs at various levels such as sensor level, feature level, score level, and decision level. Early integration strategies should be more effective than later integration strategies. Decision-level fusion is the simplest form of aggregation and requires the least amount of information. Score-level fusion requires the normalisation of feature vectors. Compared with score level and decision fusion level, sensor level and feature level have the best information input, but research data shows that little work has been done on sensor level and feature level fusion. The primary aim of this research work is to integrate randomly selected biometric modalities at the sensor level and feature level.

Two design methods have been developed for multimodal biometric systems. The first approach uses combining two randomly selected biometrics at the sensor level, while the second approach uses combining two randomly selected biometrics at the feature level. Two biometric options randomly reduce the search area during verification. Both methods reduce time and increase the accuracy of biometric systems. Additionally, randomly selected biometrics will enable the system to interact with live users.

The various sensors are used to acquire face, fingerprint, and palmprint images. Most researchers described multiresolution image fusion of the same subject which is collected from multiple sensors. The presented approach for image fusion of different subjects are collected from different sensors.

This book presents sensor-level fusion of any two randomly selected biometrics using the DWT algorithm. The second-level decomposition of DWT is used for fusion to make compatible raw data and to reduce the dimension of the input image. To obtain a feature vector, a combination of only low-frequency features and a combination of low-frequency and high-frequency features after applying DWT are used.

Low-frequency features provide positive transform values and other high-frequency elements, such as vertical, horizontal, and diagonal detail of transform values, which provide fluctuating transform values around zeros. More changes in brightness mean changes that represent a larger change in key features in the image, such as edges, lines, and borders. Therefore, the larger value of the two DWT coefficients at each point is selected to combine the two images. For fusing any two randomly selected modalities, sensor-level fusion with a combination of low and high frequency (LL, LH, HL, and HH) obtained the best results as compared to only combined low-frequency features (LL).

Incompatibility of the information content or heterogeneity is responsible for fusion at the sensor-level fusion. For example, in a face and hand geometry multi-modal system, fusion of raw images may not be possible due to the heterogeneity of features obtained from face and hand geometry. So sensor-level fusion of hand geometry features with face, palm, and fingerprint is not possible due to the heterogeneity of features.

For feature-level fusion, a unique feature vector is created from the textural information of two randomly selected biometrics. This integration resolves the differences between inter-class and intra-class variability and helps make decisions based on combined raw data, providing a small sample size and the ability to prevent attacks. Feature-level fusion is achieved with block variance features and contourlet transform features. For feature-level fusion, contourlet transform features attain more accuracy compared to block features as contourlet transform decompose the image into low- and high-frequency contourlet coefficients with different scales and various angles. Contourlet transform is capable of extracting unique textural patterns from palmprint, face, and fingerprint images. Results show that the fusion using Sum features gives a better recognition performance. Minimum and concatenated are very simple but give average performance. Maximum gives a poorly performance. The Sum rule performs best because it is less sensitive to errors in probability distributions and in this rule genuine and imposter classes are not ambiguous, and hence feature level strongly enhance prior probabilities. Feature-level fusion of two modalities adopting the Sum of feature vector leads to the best result as compared to maximum, minimum, and concatenation features.

The face, fingerprint, palmprint, and hand modality contain heterogeneous feature vectors. Min_Max normalising technique is used to modify the location and scale parameter of hand geometry modalities to transform the values into a common domain before concatenating them to form a single one. After concatenation of the feature vector in feature-level fusion, the size of the feature vector becomes very high. To reduce the size of feature vector, a linear discriminate analysis is used.

In the present book, random selection of biometric traits is used to ensure that a live person is indeed present at the point of data acquisition. The random subset is limited to fingerprint, palmprint, face, and hand geometry only.

8.2 CONCLUSIONS

In this hyper-connected world, where concerns about identity fraud and national security are increasing, identity management systems have become increasingly important. Biometric systems provide greater security and user convenience than traditional authentication methods. Multimodal systems, if well designed, can increase the matching accuracy of the identification process as they can combine evidence from different biometrics, increasing population coverage, and preventing spoofing attacks.

This book presents a new method for random selection of biometrics to improve the identification accuracy of multimodal biometric systems. Two design methods have been developed for multimodal biometric systems. The first approach uses combining two randomly selected biometrics at the sensor level, while the second approach uses combining two randomly selected biometrics at the feature level. Two randomly selected biometric approaches reduce the search space at the time of identification. Both methods reduce time and increase the accuracy of biometric systems. Additionally, randomly selected biometrics will enable the system to interact with live users.

The four biometrics used are face, fingerprint, palm print, and fingerprint. It appears that each biological system has its own advantages and limitations and that no system can meet all requirements such as efficiency, effectiveness, and price at the same time. Therefore, we can say that there is no universal biometric authentication; The search for the perfect biometric continues. Additionally, the choice of biometric feature depends on the specific application the feature is supposed to serve.

The results outperform other state-of-the-art methods using the feature-level fusions option. The ROC curve in terms of FAR and FRR shows the performance of our contourlet transform method for feature extraction.

In conclusion, this book proposes and describes a new method for multimodal biometric fusion using a random selection of biometric features, encompassing a novel feature extraction technique, a new sensor-level framework, and feature-level fusion. The feature-level fusion strategy using palmprint and fingerprint achieved the best recognition rate of 99.98%, and the ROC curve showed that the system achieved 100% GAR

Conclusions are drawn based on the research work carried out are as below

- The experimental results reveal that contourlet transform is the best modality-specific feature extraction algorithms for fingerprint, face, and palmprint, respectively.
- Results show that for hand geometry using 21 features for identification, accuracy is higher as compared to 12 features used for identification.
- Reported results show that the non-universal problem has been solved because multiple modalities of face, fingerprints, palmprints, and hand geometry can be made adequately for a high population compared to the limitations of unimodal systems.
- Reported results show that it would be difficult for an intruder to simultaneously mimic the finger, face, palm, and hand geometry of a legitimate person because subjects that do not match one modality may use another modality (e.g., manual workers, cuts their bodies, and bruises their fingers).
- Presented results justify that the integration of different biometric traits is possible at sensor-level fusion.
- Reported results reveal that a combination of low and high frequency obtained the best results as compared to only combination of low-frequency features after applying DWT for sensor-level fusion.
- Reported results reveal that sensor-level fusion of hand geometry features with face, palm, and fingerprint is not possible due to the heterogeneity of features.
- Results show that for feature-level fusion, contourlet transform features with LDA for dimension reduction attain more accuracy compared to block variance features.
- Experimental results show that selective selection of biometric features increases the accuracy of various types of biometric identification. Additionally, randomly selected biometrics will enable the system to interact with live users.

The main contribution of this research to the community is that in many sectors this work is directly applicable, for example, commercial applications, government applications, and forensic applications.

8.3   FUTURE SCOPE

Based on the work presented in this book, there are several possible investigations on future work that can be initiated. Feature-level fusion in multimodal biometrics can be extended by various techniques in extraction and merging.

The proposed feature extraction and fusion method is designed to process greyscale images. Convert the input image to greyscale and extract important information from the image. The framework can be extended in the future to process colour images that may contain additional information. The combination of information from the red, green, and blue colours of an image can create better fused feature vectors with richer information than a grey image.

This work can be extended to other biometrics such as iris, DNA, and gait. Similar techniques and biometric pattern extraction techniques are discussed in this article. However, other biometric materials (such as DNA and gait) require more methods to transform features into a form compatible with the fusion process, such as creating a statistical classification of points extracted from biometric images. New feature vectors resulting from such transformations can be used in the fusion process and thus recognition performance can be improved.

Min_Max normalisation technique is used for normalisation of hand geometry features. Other different techniques such as tanh, decimal, med, and z-score can be used.

For Product Safety Concerns and Information please contact our EU
representative GPSR@taylorandfrancis.com
Taylor & Francis Verlag GmbH, Kaufingerstraße 24, 80331 München, Germany